Lecture Notes in Computer Scie

Commenced Publication in 1973
Founding and Former Series Editors:
Gerhard Goos, Juris Hartmanis, and Jan van Leeuwen

Editorial Board

Barbara Caputo   Henning Müller
Tanveer Syeda-Mahmood   James S. Duncan
Fei Wang   Jayashree Kalpathy-Cramer (Eds.)

Medical Content-Based Retrieval for Clinical Decision Support

First MICCAI International Workshop, MCBR-CDS 2009
London, UK, September 20, 2009
Revised Selected Papers

 Springer

Volume Editors

Barbara Caputo
IDIAP Research Institute
1920 Martigny, Switzerland
E-mail: bcaputo@idiap.ch

Henning Müller
University of Applied Sciences Western Switzerland
3960 Sierre, Switzerland
E-mail: henning.mueller@hevs.ch

Tanveer Syeda-Mahmood
Fei Wang
IBM Almaden Research Center
San Jose, CA 95120, USA
E-mail: {stf,wangfe}@almaden.ibm.com

James S. Duncan
Yale University, New Haven, CT 06511, USA
E-mail: james.duncan@yale.edu

Jayashree Kalpathy-Cramer
Oregon Health and Science University
Portland, OR 97239-3098, USA
E-mail: kalpathy@ohsu.edu

Library of Congress Control Number: 2009943931

CR Subject Classification (1998): J.3, I.5, H.2.8, I.4, H.3

LNCS Sublibrary: SL 6 – Image Processing, Computer Vision,
Pattern Recognition, and Graphics

ISSN   0302-9743
ISBN-10   3-642-11768-6 Springer Berlin Heidelberg New York
ISBN-13   978-3-642-11768-8 Springer Berlin Heidelberg New York

springer.com

© Springer-Verlag Berlin Heidelberg 2010
Printed in Germany

Typesetting: Camera-ready by author, data conversion by Scientific Publishing Services, Chennai, India
Printed on acid-free paper   SPIN: 12796724   06/3180   5 4 3 2 1 0

Preface

We are pleased to present this set of peer-reviewed papers from the first
MICCAI Workshop on Medical Content-Based Retrieval for Clinical Decision
Support. The MICCAI conference has been the flagship conference for the medical imaging community reflecting the state of the art in techniques of segmentation, registration, and robotic surgery. Yet, the transfer of these techniques to
clinical practice is rarely discussed in the MICCAI conference. To address this
gap, we proposed to hold this workshop with MICCAI in London in September
2009. The goal of the workshop was to show the application of content-based
retrieval in clinical decision support. With advances in electronic patient record
systems, a large number of pre-diagnosed patient data sets are now becoming available. These data sets are often multimodal consisting of images (x-ray,
CT, MRI), videos and other time series, and textual data (free text reports and
structured clinical data). Analyzing these multimodal sources for disease-specific
information across patients can reveal important similarities between patients
and hence their underlying diseases and potential treatments. Researchers are
now beginning to use techniques of content-based retrieval to search for disease-
specific information in modalities to find supporting evidence for a disease or
to automatically learn associations of symptoms and diseases. Benchmarking
frameworks such as ImageCLEF (Image retrieval track in the Cross-Language
Evaluation Forum) have expanded over the past five years to include large medical image collections for testing various algorithms for medical image retrieval
and classification. This has made possible comparisons of several techniques for
visual, textual, and mixed medical information retrieval as well as for visual
classification of medical data based on the same data and tasks. The goal of the
workshop was to bring together researchers in this community who are inves-
tigating different aspects of this problem ranging from content-based retrieval
techniques to their use in clinical practice.

A total of ten papers were accepted for oral presentation at the workshop
after a rigorous double-blind peer-review process in which each paper was reviewed by at least three members of the Program Committee, drawn from re-
puted researchers in the domains medical imaging and clinical decision support.
In addition to the peer-reviewed papers that appear in these proceedings, there
were two invited talks by Dimitris Metaxas from Rutgers University, and Scott
Adelman from Kaiser Permanente. Dimitri's talk emphasized the need for modeling and statistical machine learning by illustrating it in three domains, namely,
LV (left ventricle) modeling in cardiac echo videos, ground glass segmentation in
lung nodule detection, and the analysis of changes in cellular appearance with
growth. Scott Adelman, on the other hand, presented a clinical practitioner's
perspective emphasizing the ease-of-use needed for decision support systems and
meaningful measurements to be made for clinical decision support.

With 30 minutes allotted to each presentation, there was sufficient time to discuss techniques and improvements in details. The accepted papers were mainly

divided into three categories, namely, medical image retrieval, clinical decision making, and multimodal fusion techniques. They appear in the same order in these proceedings. Specifically, the first paper by Andre et al. describes the application of endo-microscopic image retrieval using spatial and temporal features based on an automatically created vocabulary of visual words. Next, Ballerini et al. describe the classification of dermatologic stereoscopic images into benign and malignant tumors. Depeursinge et al. present work on interstitial lung diseases by working with 3D images of the lung taking into account the 3D texture of the volumes. Agarwal et al. describe retrieval of multidimensional tomography data to detect Alzheimer's disease.

The second session dealt with clinical decision-making and diagnosis support. Köktas et al. use a statistical analysis of gait data to aid clinical decision-making. The masses in mammograms were analyzed and used for image retrieval to aid diagnosis in the work of Zhang et al. Tao et al. describe an automatic annotation of general radiographs using robust learning techniques. The third and last session of the workshop dealt with multimodal fusion. Duchesne et al. again work on Alzheimer's disease, this time using a learning-based approach. Unay et al. presented a study on attribute selection for learning the automatic annotation of x-ray images using the data of the ImageCLEF benchmark on radiograph annotation that was used by two other presentations in the workshop. The last paper by Anatani et al. presents a multi-modal query expansion technique based on local image parts or regions to improve retrieval quality. This takes into account the importance of textual information in addition to the predominantly visual approaches that were discussed in the workshop.

Feedback to the presentations at the workshop was very positive and many discussions among the participants were fostered after the presentations, hopefully leading to many ideas and improved applications for the future. Although the modalities and the anatomical areas investigated varied, there was a common thread in all these papers in the use of learning methods and the need for validation with large annotated data sets.

We wish to thank the authors of all the submitted papers for their interest in the workshop. We also wish to thank the members of the Program Committee and the reviewers who assisted them, for reviewing the papers and providing valuable comments and recommendations to the authors. Finally, our thanks to Alison Noble and Daniel Elson for allowing us to organize this workshop at MICCAI 2009 and taking care of the many onsite organizational details.

We hope that you will find the papers in this proceedings volume interesting reading that reflects the state of the art of this field. We also hope that it will spark similar workshops at future MICCAI conferences.

September 2009   Tanveer Syeda-Mahmood
                                                           James Duncan
                                                          Henning Müller
                                                               Fei Wang
                                                         Barbara Caputo
                                                 Jayashree Kalpathy-Cramer

Organization

MCBR-CDS 2009 was organized as a satellite event of the 12th International Conference on Medical Image Computing and Computer-Assisted Intervention, that was held during September 20–24, 2009 in London, UK.

General Chairs

Tanveer Syeda-Mahmood	IBM Almaden Research Center, USA
James Duncan	Yale University, USA

Program Chairs

Henning Müller	University Hospital and University of Geneva, Switzerland
Fei Wang	IBM Almaden Research Center, USA

Publicity Chair

Jayashree Kalpathy-Cramer	Oregon Health and Science University, USA

Media Chair

Barbara Caputo	Idiap Research Institute, Switzerland

Program Committee

Amir Amini	University of Louisville, USA
Barbara Andre	INRIA, France
Sameer Antani	National Library of Medicine, USA
Nicholas Ayache	INRIA, France
David Beymer	IBM Almaden Research Center, USA
Albert Chung	Hong Kong University of Science and Technology
Dorin Comaniciu	SIEMENS, USA
Alejandro F. Frangi	Universitat Pompeu Fabra, Spain
Guido Gerig	UNC, USA
Hayit Greenspan	Tel Aviv University, Israel
Ghassan Hamarneh	Simon Fraser University, Canada
Nico Karssemeijer	Radboud University Nijmegen, The Netherlands

Ron Kikinis	M.D. Harvard, USA
Richard Leahy	USC, USA
Yanxi Liu	UPENN, USA
Martin London	M.D. UCSF, USA
Peter Macfarlane	University of Glasgow, UK
Sandy Napel	Stanford University, USA
Daniel Racoceanu	National University of Singapore, Singapore
Linda Shapiro	University of Washington, USA
Milan Sonka	University of Iowa, USA
Hemant Tagare	Yale, USA
Chris Taylor	University of Manchester, UK
Russ Taylor	JHU, USA
Joe Terdiman	M.D. Kaiser Permanente Division of Research, USA
Tatiana Tommasi	IDIAP Research Institute, Switzerland
Agma Traina	Sao Paulo University, Brazil
Max Viergewer	Utrecht University, The Netherlands
Pingkun Yan	Philips Research, USA

Sponsoring Institutions

IBM Almaden Research Center, San Jose, CA, USA

Table of Contents

Medical Image Retrieval

Clinical Decision Making

Multimodal Fusion

Overview of the First Workshop on Medical Content–Based Retrieval for Clinical Decision Support at MICCAI 2009

Henning Müller[1,2], Jayashree Kalpathy–Cramer[3], Barbara Caputo[4],
Tanveer Syeda-Mahmood[5], and Fei Wang[5]

[1] Geneva University Hospitals and University of Geneva, Switzerland
[2] University of Applied Sciences Western Switzerland (HES–SO), Sierre, Switzerland
[3] Oregon Health and Sciences Univeristy (OHSU), Portland, OR, USA
[4] IDIAP research center, Martigny, Switzerland
[5] IBM Almaden research center, San Jose, CA, USA

Abstract. In this paper, we provide an overview of the first workshop on Medical Content–Based Retrieval for Clinical Decision Support (MCBR–CDS), which was held in conjunction with the Medical Image Computing and Computer Assisted Intervention (MICCAI) conference in 2009 in London, UK. The goal of the workshop was to bring together researchers from diverse communities including medical image analyses, text and image retrieval, data mining, and machine learning to discuss new techniques for multimodal image retrieval and the use of images in clinical decision support. We discuss the motivation for this workshop, provide details about the organization and participation, discuss the current state–of–the–art in clinical image retrieval and the use of images for clinical decision support. We conclude with open issues and challenges that lie ahead for the domain of medical content–based retrieval.

1 Introduction

Diagnostic decision making (using images and other clinical data) is still very much an art for many physicians in their practices today due to a lack of quantitative tools and measurements [1]. Traditionally, decision making has involved using evidence provided by the patient's data coupled with a physician's a priori experience of a limited number of similar cases [2]. With advances in electronic patient record systems and digital imaging, a large number of pre–diagnosed patient data sets are now becoming available [3]. These datasets are often multimodal consisting of images (X–ray, CT – Computed Tomography, MRI – Magnetic Resonance Imaging), videos and other time series, and textual data (free text reports and structured clinical data). Analyzing these multimodal sources for disease–specific information across patients can reveal important similarities between patients and hence their underlying diseases and potential treatments. Researchers are now beginning to use techniques of content–based retrieval to search for disease–specific information in imaging modalities to find supporting

B. Caputo et al. (Eds.): MCBR-CDS 2009, LNCS 5853, pp. 1–17, 2010.

evidence for a disease or to automatically learn associations of symptoms and diseases [4] although already proposed over ten years ago [5,6]).

The diversity in medical imaging exams has risen enormously in the past 20 years as have the data amounts (The Geneva Hospital's Radiology department alone produced on average 114'000 images per day in 2009). Reading and interpreting multidimensional exams such as 4D data of a beating heart without computer support is extremely hard and requires much experience. At the same time all images are becoming available to clinicians via the electronic patient record [7]. This makes them available to potentially less experienced clinicians who relied on radiology reports beforehand and increases the risk of misinterpretations.

Benchmarking frameworks such as ImageCLEF[1] (Image retrieval track in the Cross–Language Evaluation Forum) have expanded over the past seven years to include large medical image collections for testing various algorithms for medical image retrieval [8,9,10]. This has made comparisons of several techniques for visual, textual, and mixed medical information retrieval as well as for visual classification of medical data possible based on the same data and tasks.

Image databases have also become available through several means such as the NCI[2] (National Cancer Institutes) and ADNI[3] (Alzheimer's Disease Neuroimaging Initiative). This lowers the entry burden to medical image analysis and should help to apply state–of–the–art techniques to medical imaging. Many open access journals such as BioMed Central[4] or Hindawi also make large amounts of the medical literature available that can then be indexed in tools such as ImageFinder[5] or MedSearch[6]. This search can include search for images by text and by visual means. Another tool that indexes openly accessible articles for the journals of the ARRS (American Roentgen Ray Society) is GoldMiner[7] [11]. All these tools and data can help to make the right information available to the right people at the right time to support healthcare applications including the use of visual data.

The first workshop on Medical Content–Based Retrieval for Clinical Decision Support (MCBR–CDS) was held at MICCAI (Medical Image Computing and Computer Assisted Interventions) 2009 in London, United Kingdom. The goal of the workshop was to bring together researchers from diverse communities including medical image analysis, text and image retrieval, data mining, and machine learning to discuss new techniques for multimodal image retrieval and the use of images in clinical decision support.

Content–based visual, textual, and multimodal information retrieval have been some of the promising techniques to help better manage the extremely

---

[1] http://www.imageclef.org/

[2] http://www.cancer.gov/

[3] http://www.adni-info.org/

[4] http://www.biomedcentral.com

[5] http://krauthammerlab.med.yale.edu/imagefinder/

[6] http://medgift.unige.ch:8080/MedSearch/faces/Search.jsp

[7] http://goldminer.arrs.org/

large amounts of visual information currently produced and used in most medical institutions around the world. As the topic has traditionally been close to rather applied research it has not yet been a primary topic at the MICCAI conference that rather concentrates on theoretically sound techniques in medical image processing. Still, to manage the increasingly large image archives in medical institutions, also the more theoretical researchers require to find the right images and use larger data sets, and on the other hand it is also time that image retrieval adopts some of the newer techniques of medical image processing, so bringing together the communities sounds like a logical step.

Submissions were proposed in the following principle areas of interests:

- data mining of multimodal medical data,
- machine learning of disease correlations from mining multimodal data,
- algorithms for indexing and retrieval of data from multimodal medical databases,
- disease model–building and clinical decision support systems based on multimodal analysis,
- practical applications of clinical decision support using multimodal data retrieval or analysis,
- algorithms for medical image retrieval or classification using the ImageCLEF collection.

A large variety of techniques were finally being submitted to the workshop. A selection of ten papers was taken that were presented orally at the workshop. Two high–quality invited speakers also presented their view on image analysis and retrieval for diagnosis support to round up the workshop.

2 Organization

MCBR–CDS was organized by an international set of researchers from the image retrieval, data mining and clinical decision support areas. The main organization was shared between Europe and the United States. The preface of these proceedings gives an overview of the organizers of the workshop and their roles in the process.

A sufficiently high number of reviewers were enlisted to help ensure the quality of presentations at the workshop. A total of over 30 experts from almost 20 countries helped in the review process.

3 Highlights of the Presentations

The workshop attracted researchers from industry and academia and from a multitude of domains, from computer science and imaging informatics to more clinically oriented groups. A total of 16 papers were submitted to the workshop. The double–blind review process included at least three external reviewers for each of the papers among the scientific community of over 30 international experts. All papers were then reread by the conference chairs and ranked based on the external reviews followed by a comparison of similarly scored papers. The ten best papers (60%) were chosen for oral presentation at the workshop using this process while the remaining papers were rejected. There was a good mix of papers between the image retrieval and clinical decision support domains as well as papers from a variety of clinical domains and using several imaging modalities. The discussions in the breaks and after the workshop underlined the interest of the presentations and the quality of the papers chosen to be presented at the workshop.

This section first presents an overview of the invited presentations and then goes into details into the three blocks of papers presented at the workshop on the topics image retrieval, clinical decision support and image annotation.

3.1   Invited Presentations

Two high quality key note presentations were given at the workshop to present external views on image retrieval and clinical decision support.

First, Dimitris Metaxas presented three projects of his research group in the context of image retrieval. The application domains from cardiac imaging to lung nodule detection, and combining sources for classification of pathology slides were presented. The importance of using high–quality data sets was highlighted to evaluate tools and algorithms on real, clinical, and thus naturally noisy data with all the difficulties that this implies. It was also shown that a variety of supporting techniques such as image segmentation in 3D and 4D are necessary for really detecting abnormal structures in the haystack of data that is often produced in medicine. Another important combination of sources that was presented was the classification of pathology slides using visual features and clinical data at the same time.

The second invited speaker was the clinician Scott Adelmann of Kaiser Permanente, the second largest hospital group in the United States. He first presented the challenges he sees in managing the currently 700 TB of image data stored by the group in connection with all the other clinical data . One of the biggest problems is the lack of structured data and thus a high–quality input for computerized tools to treat and interpret the data stored in past cases. He then presented several potential applications of image analysis and retrieval and their benefits. He clearly formulated his expectations to the image retrieval and multimodal analysis community in developing applications that are usable in practice and that can be evaluated in clinical settings.

3.2   Image Retrieval

The first set of papers describes applications of image retrieval in a variety of domains from microscopy, to CT images of the lung, MRI images of the brain and photographs of the skin, plus the use of visual and textual means combined for retrieval on the varied ImageCLEF database containing images from the scientific literature.

In [12], Andre et al. describe an application of endomicroscopic image retrieval using spatial and temporal features. They were motivated by the challenge of retrieving similar images using probe–based confocal laser endomicroscopy (pCLE), a recent imaging modality. Given a new image, they wanted to retrieve semantically similar images from a database of images annotated with expert diagnoses with the goal of aiding a physician trying to establish a diagnosis for the image. They extended the standard visual bag–of–words approach to content–based image retrieval (CBIR) by incorporating spatial and temporal information. Instead of using only salient points, they observed improved results by sampling more densely across the image. The incorporated spatial information using a co–occurrence of visual words. The temporal dimension was addressed by incorporating successive frames, a common approach in video retrieval. They also used image mosaicing to effectively increase the field of view (FOV). Using a leave–n–out cross validation technique, the authors demonstrated the efficacy of this approach by providing encouraging results on a small database of about a 1000 manually curated images. Particularly the visualization of the features of the salient points helped to understand this technique that is very often seen as a black box and thus hard to interpret.

Ballerini et al. [13] presented query–by–example image retrieval for non–melanoma skin lesions. Dermatology, a visually oriented domain, has long been popular for computer–based image analysis and automated detection systems but relatively few CBIR systems have been described in the literature. The authors reiterate the need of CBIR systems in dermatology by highlighting the importance for being able to retrieve images that might be similar in appearance to cases on hand but having different diagnoses. This would be a useful tool for a clinician in identifying differential diagnoses. Their system focussed on five non–melanoma types of lesions including Actinic Keratosis, Basal Cell Carcinoma, Melanocytic Nevus, Squamous Cell Carcinoma and Seborrhoec Keratosis. They extracted a variety of color and texture features and used a genetic algorithm for the feature selection. Finally, they evualuted their system using precision recall curves. A particular emphasis was put on the database that contains fairly different lesions and different imaging techniques than most other computer–based tools to aid diagnosis in dermatology.

Depeursinge et al. [14] discussed their work in creating a computer–aided diagnosis (CAD) system that retrieves similar cases for interstitial lung diseases (ILDs) using 3–D high resolution computerized tomography images. The goal again is to assist clinicians, in this case emergency radiologists who often need to decide very quickly and are not experts in all application domains, in the process of establishing a diagnosis. They view automated tissue classication in the lung as being complementary to case–based retrieval from a computational as well as a user–centric standpoint. They began with a semi–automated segmentation of the lung, where only a single seed point is required. This was followed by texture based categorization of lung tissue using grey level histograms and wavelets as features, and a support vector machine classifier. Similar medical cases were then retrieved using a multimodal distance measure based on the volumes of segmented tissue groups as well as text–based clinical parameters extracted from the patient's health record including patient demographics, smoking status, laboratory results, and medical history. They compared the automated tissue segmentations with annotations performed by two expert radiologists. They achieved relatively good performance in tissue categorization and case–based retrieval by incorporating the clinical context of the patient. This is actually a critical aspect of image retrieval and decision support as the context of the patient (for instance age or smoking status) can affect the visual appearance of the image used for diagnosis. By combining the text–based clinical parameters with 3D imaging, the authors have created a helpful aid for diagnosis that is currently being tested in clinical practice.

In [15], Agarwal et al. describe a computerized image retrieval and diagnosis system for Alzheimer's disease (AD). Using the popular Alzheimer's Disease Neuroimaging Initiative (ADNI) MRI data–set, the authors have created a multilevel system for indexing and retrieving similar images based on textual or visual queries. They broach important trade–offs including retrieval versus classification, representation versus classification and representation versus retrieval in this paper. Their approach to retrieval is based on first classifying the query MRI image into one of three classes — AD, mild cognitive impairment and normal. This image classification is performed using Discrete Cosine Transform (DCT) features and an SVM classifier. Once the image has been classified, the ordered list of images returned to the user from within the class is identified. The authors provide results from an evaluation that demonstrates that the precision using the two step approach of reducing the search space first by classification and then retrieving similar images is better than retrieval using all possible images.

Rahman et al. [16] report on their results with multimodal image retrieval using the ImageCLEFmed 2008 database. They have approached the problem of retrieving relevant images from a collection of medical images and annotations using a multi–modal query expansion method that integrates both visual and textual features. Using a local feedback approach, they establish correlations between visual and text keywords. Their interactive system allows either keyword–based searches or query–by–example searches. The authors manually defined 30 local concept categories and using local color and textures features, used an SVM to train the image collection. The multi–modal similarity distance is then a weighted sum of visual and textual distances. The authors demonstrate an improved performance with the query expansion using the ImageCLEFmed 2008 database.

3.3 Clinical Decision Support

This set of papers dealt with the use of images and image retrieval techniques for clinical decision making.

The Breast Image Report and Data System (BI–RADS) [17] lexicon was developed by the American College of Radiology (ACR) to standardize the terminology in describing lesions in mammogram reports. It has been used extensively in breast cancer research for classifying mammograms and more recently, ultrasounds. In this paper, Zhang et al. [17] use an ensemble approach to classifying mammograms using the BI–RADS descriptors. Using an information–gain based approach to feature selection, they identified margin and shape to be most informative while noting that age and density could be left out without sacrificing performance. They first quantized the descriptors into coarser categories and then classified each category using the best classifier from an ensemble. The authors demonstrated that they achieved equivalent results to full fine–grained representations using the the coarse–grained descriptors of a subset of the BI–RADS features. They also noted that an ensemble learning approach outperformed the individual classifiers.

Quantitative gait analysis has been used in the diagnosis and treatment of a variety of illnesses in which the disease can have a profound impact on a patient's gait including Parkinson's disease, cerebral palsy and arthritis. The paper by Sen Köktas et al. [18] uses parameters to distinguish normal patients from those suffering from knee osteoarthritis (OA) using a set of 111 patients and 110 age–matched normal subjects. Using a commercial system, the researchers collected temporal changes of joint angles, joint moments, joint power, force ratios and time–distance parameters from four anatomical locations (pelvis, hip, knee and ankle) and in three motion planes (sagittal, coronal and traversal). For each of the 33 gait attributes, 51 samples were taken in the gait cycle, resulting in a 1653–dimensional feature vector. The number of features was first reduced using either a time–averaging or FFT technique. This was followed by a further reduction of dimensionality using the Mahalanobis distance, resulting in a final feature vector dimension of about 50. Finally, these features are used for classification using a set of linear and non–linear classifiers. The authors identified highly discriminative features using the Mahalanobis distance that corresponded well with those suggested by gait analysis experts and demonstrated good performance in the classification task using non–linear classifiers coupled with the use of time–averaging for dimensionality reduction. This paper could unfortunately not be presented at the workshop as the authors were unable to come at the very last moment.

Duchesne et al. [19] have described their approach to integrating information from various modalities including clinical, cognitive, genetic and imaging data to create a decision model to discriminate patients suffering from Alzheimer's disease from normal controls. The data are from the ADNI database include the results of neuropsychological tests, quantitative hippocampal volumes obtained from imaging, and demographic and genetic risk factors including age, gender and APOE genotype. These data were integrated in the form of a binary string allowing the use of the Hamming distance for classification. The authors reported a 99.8% classification accuracy using 10–fold cross–validation.

3.4 Automatic Image Annotation

The final set of papers deals with the concept of automatic annotation of images based on visual appearance. Both papers in this set used the data from the Automatic Annotation task of ImageCLEF [8]. This annual medical image classification challenge consists of automatically annotating an image collection of more than 2000 X–rays given a training set of about 12'000 classified images. The images are classified using the IRMA (Image Retrieval in Medical Applications) [20] scheme with hierarchical labels for imaging modality, anatomical location, body system and image view. Some of the challenging aspects of this task include the highly unbalanced class memberships, significant intra–class visual dissimilarity as wells and inter–class similarity. The scoring system for the task was set up to reduce guessing when in doubt by penalizing an incorrect class more than an "unknown" class and penalizing mistakes lower down in the hierarchy less than errors closer to the top. The goals of this hierarchical weighting scheme was to force groups to add a confidence into the classification evaluation and give groups with good confidence score a better result. In 2009, the task was also made more difficult in that the distribution of the number of elements per class in training and testing data was deliberately not the same and even images from classes not occurring in the training data could be part of the test data.

Unay et al. [21] discuss their results of their participation in the Image-CLEFmed automatic annotation task. The main contribution of this paper was demonstrating that PCA–based local binary patterns used for the SVM classifier performed almost as well as the complete feature vector set, thereby enabling the use of a smaller dimensional feature vector and reducing computation time. This resulted in a 5–fold improvement in processing time and storage space requirements.

In [22], the authors propose a novel learning–based algorithm for medical image annotation that utilizes robust aggregation of learned local appearance evidences. This approach was applied to the task of automatically distinguishing the posteroanterior/anteroposterior (PA–AP) from the lateral (LAT) views of chest radiographs with the goal of integrating this as a post–processing module for computer–aided detection systems for both an in–house database as well as the ImageCLEF automatic annotation collection. The authors begin by demonstrating the within class variability found in appearance of radiographs of the chest and pelvis. The algorithm to identify the view of chest radiographs starts with the detection of landmarks using simultaneous feature selection and classification at different scales. Next, these landmarks are filtered using a sparse configuration algorithm to eliminate inconsistent findings. Finally a reasoning module identifies the final image class using these remaining landmarks. They demonstrate superior results in distinguishing PA–AP from LAT views for chest radiographs. An evaluation using a subset of the ImageCLEFmed classification collection was also performed using the most frequent classes of the database. The approach shows very good performance on a small number of classes with reasonably large difference between the classes. The scalability to a large number of classes was not attempted.

3.5   General Remarks on the Workshop

The workshop attracted researchers from a variety of clinical domains working on the challenges associated with image retrieval, primarily to aid with clinical decision support. Many different imaging modalities (x–ray, CT, MRI, endoscopy, photography, dermatology, microscopy) were presented as well as many anatomic regions (head, lung, colon, brain, skin, varied anatomic regions). Some traditional domains such as MRI image retrieval of the brain and CT image analysis of the lung were present as well as general image classification of X–rays. On the other hand several new domains were described such as the micro–endoscopy system described in [12]. Whereas most traditional approaches analyzed also 3D data mainly in 2D slices there were several approaches analyzing and calculating similarity directly in the 3D space [14,15]. Time–series of images were tackled as well as combining visual features with clinical parameters.

However, there are also several shortcoming that became apparent in this workshop. Few of the approaches are developed in close clinical collaboration meaning that a real use of the systems was basically not evaluated at all. Most often, publicly available databases or parts of them were used for evaluating specific aspects for the diagnosis aid process [15,17,19]. Sometimes, own databases were created with clinicians [13] but no full clinical evaluation was attempted. In general it is very hard to compare techniques and systems as often databases and setups are different for each system making comparison impossible. The next section will go deeper into the current challenges of image retrieval as diagnosis aid.

4   Open Issues and Challenges

The papers presented at the workshop show that the domain of medical image retrieval for clinical decision support is moving forward strongly. Still, some limitations could also be identified that require to be tackled in the near future to make image retrieval a tool usable for clinical practice. Notable, these challenges are:

- currently, only very few image retrieval systems have been developed in close collaboration with clinicians and have been evaluated clinically in a real workflow;
- purely visual systems do not take into account the clinical context and a combination seems necessary for a full decision support;
- standards for evaluation including data sets, ground truth ad criteria seem necessary to really be able to compare results on the same basis and show advances in the field;
- multidimensional data sets represent by far the largest amount of data produced in hospitals at the moment, i.e. 3D and 4D data as well as combinations of modalities, which have not been used for image retrieval, yet;
- the sheer size of PACS is not treated by any image retrieval system at the moment, particularly small Matlab prototypes will not scale to several million images.

These challenges will be detailed in the following sections.

4.1   Integration of Clinicians and Clinical Evaluation

One of the biggest challenges at the moment is bringing the theoretical applications that are often performed on small data sets and with MatLab prototypes towards real clinical applications. To our knowledge only a single evaluation has so far been performed with medical image retrieval [23] showing a clear improvement of diagnosis quality with the system use, particularly for little experienced radiologists. Most often, systems are developed in computer science departments far away from clinicians and with no direct collaboration other than an exchange of data. A close collaboration with clinicians and including frequent feedback from clinicians is necessary, which can become possible through direct collaborative projects.

Another important part in collaboration with clinicians is the analysis of the behavior of clinicians for example when using images or searching for them [24,25]. Health care professionals have indicated that they would like to be able to restrict searches to a given modality, anatomy, or pathology of the image. However, the image annotations in on–line collections or teaching files do not always contain the information about the modality or anatomy. On the other hand, purely visual systems are not believed to be mature enough for image retrieval for images with specific pathological findings, especially for image collections containing a variety of image modalities and pathologies. One thing mentioned was also the search for visually similar images but with different diagnosis, to illustrate teaching and also for differential diagnoses. One big problem with visual retrieval currently is that the formulation of information needs with visual means is far from easy.

Another important aspect for medical image retrieval is the notion of relevance in medical image search. This is somewhat researched for still images but for 3D or 4D data sets this is not clear. Past tests have also shown that this is person–dependent, task dependent, and depends strongly on the knowledge of the searching person in a particular domain [9].

Another way to convince clinicians is to have a clear proof of retrieval quality as few people would want to work with systems that can not show a certain quality level. To show such performance, standard data sets are extremely important [26] and also a methodology to evaluate several systems based on the same means.

4.2   Multimodal Data Treatment and Information Fusion

Purely visual techniques may not be sufficient for most clinical applications. In medicine, visual information taken alone is less meaningful than the same images viewed in the context of the patient and the clinical environment. We believe that pure CBIR methods in medicine have not lived up to expectations due to their inability to incorporate context. No medical doctor would diagnose based on images, only, as the context carries much of the necessary information to interpret images. Image retrieval in medicine needs to evolve from purely visual retrieval to a more holistic, case–based approach that incorporates various multimedia data sources. These include multiple images, free text, structured data, as well as external knowledge sources and ontologies.

The semantic gap poses one of the major challenges in creating a useful image retrieval engine. Smeulders [27] identified the *semantic gap* as the lack of coincidence between the information that one can automatically extract from

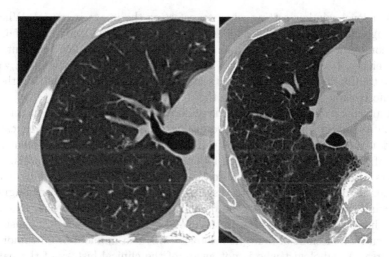

Fig. 1. Comparing the healthy tissue of a 25–year-old and an 88-year–old person shows the important differences in grey level and texture

the visual data and the interpretation that the same data have for a given user in a given situation. In medical images, the semantic gap can manifest itself as a difference between the image and the interpretation of the image by the medical doctor including anamnesis, lab results, and potentially other exams. The same image may be interpreted differently depending on the medical doctor, his training, expertise, experience, and the context of the image acquisition and the patient. Such coincidences between content and contextual data have already been described in the non–medical field in [28] as well.

Effective clinical image retrieval systems can be used as a diagnostic aid. By allowing clinicians to view similar images contextually, they receive assistance in the diagnostic decision–making process by accessing knowledge of older cases. When being pro–active in this process missing data such as lacks in the anamnesis can be pointed out by the system and the clinician can directly ask the questions with the highest clinical information gain to the patient or order the corresponding lab examinations, as proposed by a computerized decision aid. This of course requires much more knowledge about a particular domain and the interrelations of the clinical data.

Examples for the importance of combining the textual and visual data are manifold. Figure 1 shows as an example the healthy lung of a 25–year–old and an 88-year–old. The lung of the 88–year–old shows several pre–fibrotic lesions and has a slightly altered grey value. Inverting age on the image of the 88–year–old would mean that the persons is not healthy but has a severe problem. Another example is the importance of the goal of the imaging study as it provides the context in which the image is to be viewed. CT images have a high dynamic range. The window/level settings must be set appropriately to provide detail and contrast for the organ of interest in the imaging study. Often, images are stored in JPEG for teaching and conference presentations and also in this case the right level/window setting when transferring the image is crucial. Whereas CT images usually have 1000–4000 grey levels, jpeg images only have 256, and most computer screens do not manage to show more than 256 different grey levels, either. Looking at chest CTs of the mediastinum or of the lungs would require totally different level window settings than looking at the lung, although the exactly same regions is show on the image.

Yet another example deals with the ability to incorporate patient history into the context in which the images are evaluated. In patients with lung cancer, radiation therapy is often delivered to the chest as part of the treatment plan. Many of these patients develop lung inflammation, known as pneumonitis. Some patients also develop radiation fibrosis, a scarring of the lungs. This can be mistaken for other interstitial lung diseases if the context of the patient is ignored in viewing subsequent scans of the chest. There are numerous other examples where the role of context is vital in the use of imaging studies for diagnosis and treatment. The lesions of multiple sclerosis (MS) can mimic a brain tumor and vice versa. A radiologist who is not aware of the clinical history of the patient as having MS can misdiagnose a suspicious lesion on an MRI.

All these examples underline that images can basically not be viewed correctly without clinical information and albeit this, most of the medical image retrieval systems currently ignore clinical data other than images totally.

4.3   Treating Extremely Large Databases

In the same way as for general Internet search engines one of the most important aspects for medical image retrieval systems is to be as complete as possible and as large as possible. Medical image repositories have multiplied over recent years with tools such as MyPACS[8] and MDPixx[9]. Even a standard to interconnect digital teaching files exists with MIRC[10] (Medical Imaging Resource Center). With the scientific literature another large body of knowledge that includes many medical images has become accessible [29] and is increasingly integrated with visual retrieval systems. Several web interfaces such as Goldminer[11] allow access to many hundred thousand images and these numbers are very likely to increase strongly and quickly making available for information search ever larger amounts of medical knowledge including images.

The daily image production at the Geneva University Hospitals' radiology department (see Figure 2) also shows that internal data sets have grown exponentially and are continuing to grow at these rates. Multi–slice scanners and combinations of modalities such as PET/CT are some of the largest data producers. Large University hospitals often produce in access of 100 GB per day. Indexing the entire PACS for image retrieval in clinical routine has been proposed many times [30,31] but to our knowledge not a single implementation has

---

[8] http://www.mypacs.net/
[9] http://www.mdpixx.org/
[10] http://mirc.rsna.org/
[11] http://goldminer.arrs.org/

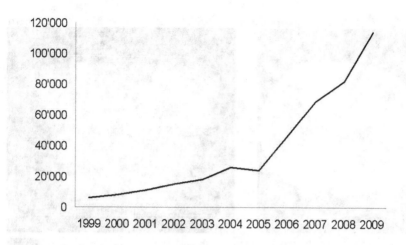

Fig. 2. The daily image production in the Radiology Department of the Geneva University hospitals has increased enormously over the past ten years

been performed up to now. Blocking parts include the legal aspects of accessing patient data but also the sheer amount of information that will require new index structures to cope with the several million images thus potentially available. Large benchmarking databases currently contain rarely more than 100'000 images, which is often already on the limit for prototypes in MatLab or systems that contain the features in main memory. Indexing and thus making accessible extremely large data sets still contains many challenges, also in reaching interactive response times.

4.4   Treating Multidimensional Data Sets

Currently, image retrieval is most often limited to single two–dimensional images. CT, MRI and also combined modalities such as PET/CT and PET/MRI are by far the largest production of data in hospitals (including time series of such data, so 4D data sets that have never been used for image retrieval, so far). Only very few systems analyze these multidimensional data sets directly.

Already viewing these data sets creates the challenging parts as humans are good in qualitative analysis but can usually not remember more than a few things at a time (from 3–7 according to the psychological literature [32,33]). This means that viewing such multidimensional data with many aspects requires much experiences and puts a stress on the clinicians by having to integrate very large amounts of data. Analyzing the data with diagnosis aid tools and highlighting potentially abnormal regions can at least reduce the stress through at least a somewhat second opinion.

Figure 3 shows a simple image retrieval system that visualizes the obtained cases directly in 3D and shows them to the clinicians. This interface is a web interface (thus easy to integrate into clinical applications) using Java3D and

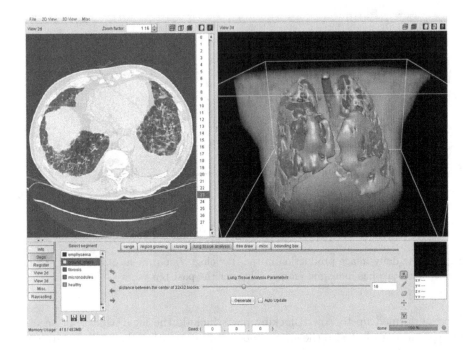

Fig. 3. An interface visualizing classified tissue in a 2D and a 3D view

particularly the YaDiV[12] (Yet another DICOM Viewer) system for visualization. The interface allows the clinicians to navigate directly in the 3D data if necessary but to concentrate on areas with abnormal parts in more details by automatically pre–classifying the entire lung tissue ahead of viewing.

5   Conclusions

This paper shows the advances that image retrieval has made over the past years through the ten presentations given at the workshop on Medical Content–Based Retrieval for Clinical Decision support. It can be seen that image retrieval is leaving the paradigm of taking image similarity from single images and that an integration of data from 3D data sets, clinical data, and also temporal data of time series has started. Standard data sets are available in some domains and are used increasingly, albeit not always in the same ways. Data sets need to contain ground truth and clear criteria of success to allow for a real comparison of techniques to measure progress.

Systems also need to get closer to clinicians and show their potential with, at least, small trials. To do so, the integration of clinical with the visual data seem necessary. Very large amounts of clinical data are available and their integration

---

[12] http://www.welfenlab.de/en/research/fields_of_research/yadiv/

into clinical applications using techniques from image retrieval seems necessary. Such large amounts are often accessible but they need to be integrated and accessible quickly.

Image retrieval in medicine needs to evolve from purely visual image retrieval to a more holistic, case–based approach that incorporates various multimedia data sources and thus the context in which the images were taken. This needs to include multiple images, free text, structured data as well as external knowledge sources and ontologies. All these data can consequently be integrated with literature databases such as Goldminer to give a clinician access to the right information (peer–reviewed literature, past cases with treatment and outcomes) at the right time and in the right format.

Acknowledgements

This work was partially funded by the Swiss National Science Foundation (FNS) under contract 205321–109304/1, the American National Science Foundation (NSF) with grant ITR–0325160, the TrebleCLEF project and Google. We would like to thank the RSNA for supplying the images of their journals Radiology and Radiographics for the ImageCLEF campaign. Thanks also to the HES–SO for funding the BeMeVIS project. A big Thank You also to the two invited speakers who helped to make the workshop a success. Thanks belong equally to IBM for supporting the workshop organization. Particular thanks also go to the MICCAI workshop organizer Daniel Elson for well organizing all logistical and practical aspect of the workshop and for remaining calm when we not respecting deadlines or forgot to respond to important mails.

References

1. Kohn, L.T., Corrigan, J.M., Donaldsen, M.S.: To Err is Human – Building a Safer Health System. National Aacademic Press, Washington (1999)
2. Aamodt, A., Plaza, E.: Case–based reasoning: Foundational issues, methodological variations, and systems approaches. Artificial Intelligence Communications 7(1), 39–59 (1994)
3. Safran, C., Bloomrosen, M., Hammond, W.E., Labkoff, S., Markel-Fox, S., Tang, P.C., Detmer, D.E.: Toward a national framework for the secondary use of health data: An american medical informatics association white paper. Methods of Information in Medicine 14, 1–9 (2007)
4. Müller, H., Michoux, N., Bandon, D., Geissbuhler, A.: A review of content–based image retrieval systems in medicine – clinical benefits and future directions. International Journal of Medical Informatics 73(1), 1–23 (2004)
5. Tagare, H.D., Jaffe, C., Duncan, J.: Medical image databases: A content–based retrieval approach. Journal of the American Medical Informatics Association 4(3), 184–198 (1997)
6. Lowe, H.J., Antipov, I., Hersh, W., Arnott Smith, C.: Towards knowledge–based retrieval of medical images. The role of semantic indexing, image content representation and knowledge–based retrieval. In: Proceedings of the Annual Symposium of the American Society for Medical Informatics (AMIA), Nashville, TN, USA, October 1998, pp. 882–886 (1998)

7. Haux, R.: Hospital information systems — past, present, future. International Journal of Medical Informatics 75, 268–281 (2005)
8. Müller, H., Kalpathy-Cramer, J., Kahn Jr., C.E., Hatt, W., Bedrick, S., Hersh, W.: Overview of the ImageCLEFmed 2008 medical image retrieval task. In: Peters, C., Giampiccolo, D., Ferro, N., Petras, V., Gonzalo, J., Peñas, A., Deselaers, T., Mandl, T., Jones, G., Kurimo, M. (eds.) Evaluating Systems for Multilingual and Multimodal Information Access – 9th Workshop of the Cross-Language Evaluation Forum. LNCS, vol. 5706, pp. 500–510. Springer, Heidelberg (2009)
9. Müller, H., Kalpathy-Cramer, J., Eggers, I., Bedrick, S., Said, R., Bakke, B., Kahn Jr., C.E., Hersh, W.: Overview of the 2009 medical image retrieval task. In: Working Notes of CLEF 2009 (Cross Language Evaluation Forum), Corfu, Greece (September 2009)
10. Müller, H., Deselaers, T., Lehmann, T., Clough, P., Kim, E., Hersh, W.: Overview of the ImageCLEFmed 2006 medical retrieval and annotation tasks. In: Working Notes of the 2006 CLEF Workshop, Alicante, Spain, Septermber (2006)
11. Kahn Jr., C.E., Thao, C.: Goldminer: A radiology image search engine. American Journal of Roentgenology 188, 1475–1478 (2008)
12. André, B., Vercauteren, T., Perchant, A., Buchner, A.M., Wallace, M.B., Ayache, N.: Introducing space and time in local feature-based endomicroscopic image retrieval. In: Müller, H. (ed.) MCBR–CDS 2009. LNCS, vol. 5853, pp. 18–30. Springer, Heidelberg (2009)
13. Ballerini, L., Fisher, R., Rees, J.: A query–by–example content–based image retrieval system of non–melanoma skin lesions. In: Caputo, B., Müller, H., Syeda Mahmood, T., Kalpathy-Cramer, J., Wang, F., Duncan, J. (eds.) MCBR–CDS 2009. LNCS, vol. 5853, pp. 31–38. Springer, Heidelberg (2009)
14. Depeursinge, A., Vargas, A., Platon, A., Geissbuhler, A., Poletti, P.A., Müller, H.: 3D case–based retrieval for interstitial lung diseases. In: Caputo, B., Müller, H., Syeda Mahmood, T., Kalpathy-Cramer, J., Wang, F., Duncan, J. (eds.) MCBR–CDS 2009. LNCS, vol. 5853, pp. 39–48. Springer, Heidelberg (2009)
15. Agarwal, M., Mostafa, J.: Image retrieval for alzheimer's disease detection. In: Caputo, B., Müller, H., Syeda Mahmood, T., Kalpathy-Cramer, J., Wang, F., Duncan, J. (eds.) MCBR–CDS 2009. LNCS, vol. 5853, pp. 49–60. Springer, Heidelberg (2009)
16. Rahman, M., Antani, S.: Multi–modal query expansion based on local analysis for medical image retrieval. In: Caputo, B., Müller, H., Syeda Mahmood, T., Kalpathy-Cramer, J., Wang, F., Duncan, J. (eds.) MCBR–CDS 2009. LNCS, vol. 5853, pp. 110–119. Springer, Heidelberg (2009)
17. Zhang, Y., Tomuro, N., Furst, J., Raicu, D.S.: Using bi–rads descriptors and ensemble learning for classifying masses in mammograms. In: Caputo, B., Müller, H., Syeda Mahmood, T., Kalpathy-Cramer, J., Wang, F., Duncan, J. (eds.) MCBR–CDS 2009. LNCS, vol. 5853, pp. 69–76. Springer, Heidelberg (2009)
18. Sen Köktas, N., Duin, R.P.W.: Statistical analysis of gait data to assist clinical decision making. In: Caputo, B., Müller, H., Syeda Mahmood, T., Kalpathy-Cramer, J., Wang, F., Duncan, J. (eds.) MCBR–CDS 2009. LNCS, vol. 5853, pp. 61–68. Springer, Heidelberg (2009)
19. Duchesne, S., Crépeault, B., Frisoni, G.: Knowledge–based discrimination in alzheimer's disease. In: Caputo, B., Müller, H., Syeda Mahmood, T., Kalpathy-Cramer, J., Wang, F., Duncan, J. (eds.) MCBR–CDS 2009. LNCS, vol. 5853, pp. 89–96. Springer, Heidelberg (2009)

20. Lehmann, T.M., Schubert, H., Keysers, D., Kohnen, M., Wein, B.B.: The IRMA code for unique classification of medical images. In: Huang, H.K., Ratib, O.M. (eds.) Medical Imaging 2003: PACS and Integrated Medical Information Systems: Design and Evaluation, San Diego, California, USA, May 2003. Proceedings of SPIE, vol. 5033, pp. 440–451 (2003)

21. Unay, D., Soldea, O., Ekin, A., Cetin, M., Ercill, A.: Automatic annotation of x–ray images: A study on attribute selection. In: Caputo, B., Müller, H., Syeda Mahmood, T., Kalpathy-Cramer, J., Wang, F., Duncan, J. (eds.) MCBR–CDS 2009. LNCS, vol. 5853, pp. 97–109. Springer, Heidelberg (2009)

22. Tao, Y., Peng, Z., Jian, B., Xuan, J., Krishnan, A., Zhou, X.S.: Robust learning based annotation of medical radiographs. In: Caputo, B., Müller, H., Syeda Mahmood, T., Kalpathy-Cramer, J., Wang, F., Duncan, J. (eds.) MCBR–CBS 2009. LNCS, vol. 5853, pp. 77–88. Springer, Heidelberg (2009)

23. Aisen, A.M., Broderick, L.S., Winer-Muram, H., Brodley, C.E., Kak, A.C., Pavlopoulou, C., Dy, J., Shyu, C.R., Marchiori, A.: Automated storage and retrieval of thin–section CT images to assist diagnosis: System description and preliminary assessment. Radiology 228(1), 265–270 (2003)

24. Hersh, W., Jensen, J., Müller, H., Gorman, P., Ruch, P.: A qualitative task analysis for developing an image retrieval test collection. In: ImageCLEF/MUSCLE workshop on image retrieval evaluation, Vienna, Austria, pp. 11–16 (2005)

25. Müller, H., Despont-Gros, C., Hersh, W., Jensen, J., Lovis, C., Geissbuhler, A.: Health care professionals' image use and search behaviour. In: Proceedings of the Medical Informatics Europe Conference (MIE 2006), Maastricht, The Netherlands. Studies in Health Technology and Informatics, pp. 24–32. IOS Press, Amsterdam (2006)

26. Vannier, M.W., Summers, R.M.: Sharing images. Radiology 228, 23–25 (2003)

27. Smeulders, A.W.M., Worring, M., Santini, S., Gupta, A., Jain, R.: Content–based image retrieval at the end of the early years. IEEE Transactions on Pattern Analysis and Machine Intelligence 22(12), 1349–1380 (2000)

28. Westerveld, T.: Image retrieval: Content versus context. In: Recherche d'Informations Assistée par Ordinateur (RIAO 2000) Computer–Assisted Information Retrieval, Paris, France, CID, April 2000, vol. 1, pp. 276–284 (2000)

29. Müller, H., Kalpathy-Cramer, J., Kahn Jr., C.E., Hersh, W.: Comparing the quality of accessing the medical literature using content–based visual and textual information retrieval. In: SPIE Medical Imaging, Orlando, Florida, USA, February 2009, vol. 7264, pp. 1–11 (2009)

30. Bueno, J.M., Chino, F., Traina, A.J.M., Traina, C.J., Azevedo-Marques, P.M.: How to add content–based image retrieval capacity into a PACS. In: Proceedings of the IEEE Symposium on Computer–Based Medical Systems (CBMS 2002), Maribor, Slovenia, pp. 321–326 (2002)

31. Qi, H., Snyder, W.E.: Content–based image retrieval in PACS. Journal of Digital Imaging 12(2), 81–83 (1999)

32. Cowan, N.: The magical number 4 in short–term memory: A reconsideration of mental storage capacity. Behavioral and Brain Sciences 24(1) (2001)

33. Miller, G.A.: The magical number seven plus or minus two: Some limits on our capacity for processing information. The Psychological Review 63, 81–97 (1956)

Introducing Space and Time in Local Feature-Based Endomicroscopic Image Retrieval

Barbara André[1,2], Tom Vercauteren[1], Aymeric Perchant[1], Anna M. Buchner[3],
Michael B. Wallace[3], and Nicholas Ayache[2]

[1] Mauna Kea Technologies (MKT), Paris, France
[2] INRIA - Asclepios Research Project, Sophia Antipolis, France
[3] Mayo Clinic, Jacksonville, Florida, USA

Abstract. Interpreting endomicroscopic images is still a significant challenge, especially since one single still image may not always contain enough information to make a robust diagnosis. To aid the physicians, we investigated some local feature-based retrieval methods that provide, given a query image, similar annotated images from a database of endomicroscopic images combined with high-level diagnosis represented as textual information. Local feature-based methods may be limited by the small field of view (FOV) of endomicroscopy and the fact that they do not take into account the spatial relationship between the local features, and the time relationship between successive images of the video sequences. To extract discriminative information over the entire image field, our proposed method collects local features in a dense manner instead of using a standard salient region detector. After the retrieval process, we introduce a verification step driven by the textual information in the database and in which spatial relationship between the local features is used. A spatial criterion is built from the co-occurence matrix of local features and used to remove outliers by thresholding on this criterion. To overcome the small-FOV problem and take advantage of the video sequence, we propose to combine image retrieval and mosaicing. Mosaicing essentially projects the temporal dimension onto a large field of view image. In this framework, videos, represented by mosaics, and single images can be retrieved with the same tools. With a leave-n-out cross-validation, our results show that taking into account the spatial relationship between local features and the temporal information of endomicroscopic videos by image mosaicing improves the retrieval accuracy.

1 Introduction

With the recent technology of probe-based confocal laser endomicroscpy (pCLE) [1], endoscopists are able to image tissues at microscopic level with a miniprobe, and in real time during ongoing procedure. However, as the acquired pCLE images are relatively new for them, the physicians are still in the process of defining a taxonomy of the pathologies in the images, for instance to differentiate benign tissues and neoplastic, i.e. pathological, tissues of colonic polyps, see Fig. 1 and 2 for an illustrative example of such images. To face this clinical challenge, a valuable aid to the physician in establishing a diagnosis would be to provide

B. Caputo et al. (Eds.): MCBR-CDS 2009, LNCS 5853, pp. 18–30, 2010.

endomicroscopic images that have a similar appearance to the image of interest and that have been previously diagnosed by expert physicians. Knowing that pathological tissue is characterized by some irregularities in the cellular and vascular architecture, we aim at retrieving texture information coupled with shape information by using local operators on pCLE images. To serve that purpose, we decided to investigate a modern method for content-based image retrieval (CBIR), the bag-of-visual words (BVW) method [2]. BVW has been successfully used in many applications of computer vision. For example, on a well-defined non medical application, by using this method on a large variety of images of natural or artificial textures, the authors of [2] obtained excellent recognition results that are close to 98%.

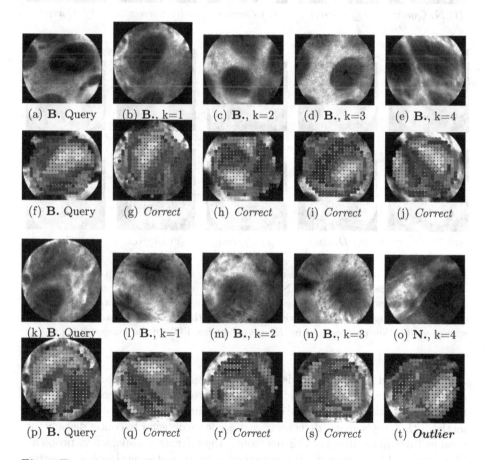

Fig. 1. Typical image retrieval results provided by our method from two benign queries. **B.** indicates Benign and **N.** Neoplastic. From left to right on each row: the queried image, and its k-NNs on the top layer, and their respective colored visual words on the bottom layer. An outlier is indicated by *Outlier* if it has been rejected by the spatial verification process, and by *Error* otherwise. FOV of the images: 240 µm.

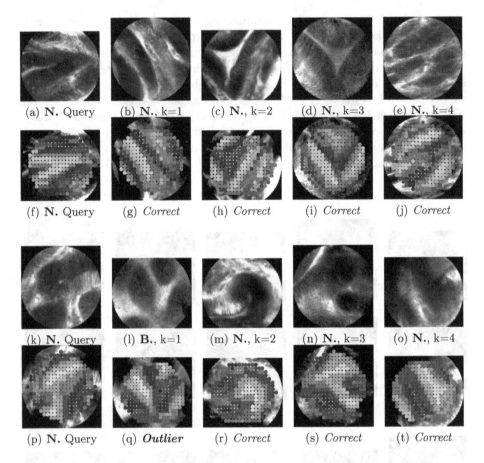

(a) **N.** Query   (b) **N.**, k=1   (c) **N.**, k=2   (d) **N.**, k=3   (e) **N.**, k=4

(f) **N.** Query   (g) *Correct*   (h) *Correct*   (i) *Correct*   (j) *Correct*

(k) **N.** Query   (l) **B.**, k=1   (m) **N.**, k=2   (n) **N.**, k=3   (o) **N.**, k=4

(p) **N.** Query   (q) ***Outlier***   (r) *Correct*   (s) *Correct*   (t) *Correct*

Fig. 2. Typical image retrieval results provided by our method from two neoplastic queries. **B.** indicates Benign and **N.** Neoplastic. From left to right on each row: the queried image, and its k-NNs on the top layer, and their respective colored visual words on the bottom layer. An outlier is indicated by ***Outlier*** if it has been rejected by the spatial verification process, and by ***Error*** otherwise. FOV of the images: 240 μm.

The standard BVW method detects salient regions in the images and extracts information only on these specific regions. However in pCLE images, the discriminative information is distributed over the entire image field. Contrary to classical methods that apply sparse detectors, we use a dense detector to collect densely the local features in the images. This overcomes the information sparseness problem. Moreover, pCLE images contain characteristic pattern at several scales, in particular the microscopic scale of individual cells and the mesoscopic scale of groups of cells. For this reason, we perform a bi-scale description of the collected image regions. Another problem is that the spatial relationship between the local features is lost in the standard BVW representation of an image, whereas the spatial organization of cells is highly discriminative in pCLE images. So we looked at measuring a statistical representation of this spatial geometry. This was achieved by exploiting the co-occurence matrix of the visual words labeling the local features in the image. After the retrieval process, we introduce the measured spatial criterion in a verification step that allows to remove outliers from the retrieved pCLE images, which are given by the most similar to queried images. Taking into account the spatial relationship between local features is the main contribution of our study, it can be used as a generic tool for many applications of CBIR. Besides, we noticed that the FOV of single still pCLE images may not be large enough for the physicians to see a characteristic global pattern and make a robust diagnosis. As this limitation cannot be solved by the standard methods, we decided to take into account the time information of pCLE video sequences by considering them as objects of interest instead of still images. More precisely, we use image mosaicing [3] to project the temporal dimension of video sequences onto a large FOV image, cf. some resulting mosaics in Fig. 4 and 5. With a leave-n-out cross-validation, classification experiments on the pCLE database serve the validation of the methodology: our method outperforms other methods taken as references, by improving the classification accuracy and by providing more relevant training images among the first retrieved images.

2   The Bag-of-Visual Words Method

As one of the most popular method for image retrieval, the BVW [2] method aims at extracting a local image description that is both efficient to use and invariant with respect to viewpoint changes, e.g., translations, rotations and scaling, and illumination changes, e.g., affine transformation of intensity. Its methodology consists in first finding and describing local features, then in quantizing them into clusters named visual words, and in representing the image by the histogram of these visual words. The BVW retrieval process can thus be decomposed into four steps: detection, description, clustering and similarity measuring, possibly followed by a classification step for image categorization.

The detection step extracts salient regions in the image, i.e. regions containing some local discriminative information. In particular, corners and blobs in the image can be detected by the sparse Harris-Hessian (H-H) operator around key-points with high responses of intensity derivatives. Other sparse detectors like the Intensity-Based Regions (IBR) and the Maximally Stable Extremal Regions (MSER) are also specialized for the extraction of blob features in the images. We refer the interested reader [4] for a survey of these detectors.

Then, each local region can be typically described by the Scale Invariant Feature Transform (SIFT) [5] descriptor. We refer the reader [2] for a survey of this and other powerful descriptors. At the description step, the SIFT descriptor computes, for each salient region, a description vector which is its gradient histogram at the optimal scale provided by the detector, the gradient orientations being normalized with respect to the principal orientation of the salient region. As a result, the image is represented in a high dimensional space by a set of SIFT description vectors that are invariant by translation, rotation and scale.

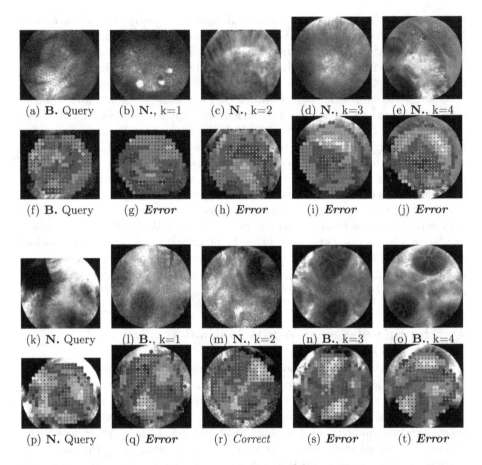

Fig. 3. Worst image retrieval results provided by our method. The benign query on the top is a rare benign variety which is not represented in the training dataset. The neoplastic query on the bottom contains on its top left corner a partially visible elongated crypt which could not be totally described. **B.** indicates Benign and **N.** Neoplastic. From left to right on each row: the queried image, and its k-NNs on the top layer, and their respective colored visual words on the bottom layer. An outlier is indicated by *Outlier* if it has been rejected by the spatial verification process, and by *Error* otherwise. FOV of the images: 240 µm.

To reduce the dimension of the description space, the clustering step, for example based on a standard K-Means, builds K clusters, i.e. K visual words, from the union of the description vector sets gathered from all the N images of the training database. Since each description vector counts for one visual word, an image is represented by a signature of size K which is its histogram of visual words, normalized by the number of its salient regions.

Given these image signatures, it is possible to define a distance between two images as the χ^2 distance [4] between their signature and to retrieve the closest training images as the most similar to the image of interest. The relevance of the similarity results can be quantified by a further classification step, for instance based on a standard nearest neighbors procedure that weights the votes of the k-nearest neighbors by the inverse of their χ^2 distance to the signature of the queried image, so that the closest images are the most determinant. Besides, performing image classification is a way to validate a new retrieval method by comparing it with other methods.

3   Including Spatial and Temporal Information

When we applied the standard BVW method on pCLE images [6], we obtained rather poor classification results and the presence of many retrieval outliers. To improve the accuracy of endomicroscopic image retrieval, we decided to include both spatial and temporal information contained in the pCLE images. While testing on pCLE images the numerous sparse detectors listed by [4], we first observed that a large number of salient regions sparsely extracted by these standard detectors do not persist between two highly correlated successive images taken from the same video. This is consistent with the rather poor results of the standard sparse detector shown in Section 4. To overcome the persistence problem and take into account all the information in the images, we use a dense detector contrarily to the standard method. Furthermore, a dense detector is relevant because local information appears to be densely distributed over the entire field of the pCLE image. The dense detector is made of overlapping disks localized on a dense regular grid, such that each disk covers a possible image pattern at microscopic level.

We also noticed that the endoscopists establish their diagnosis on pCLE images from the regularity of the cellular architecture in the colonic tissue [7], where goblet cells and crypts are both round-shaped characteristic patterns, but where a crypt has larger size than its surrounding goblet cells so it must not be recognized as the same object. In order to be sensitive to scale changes, our method looked at describing local disk regions at various scales that are not automatically computed, for example by choosing a microscopic scale for individual cell patterns and a mesoscopic scale for larger groups of cells. This leads us to represent an image by several sets of description vectors that are scale-dependent, resulting in several signatures for the image that are then concatenated into one larger signature.

This previous observation also suggests that the spatial organization of the goblet cells must be included in the retrieval process because it is substantial to differentiate benign tissues from neoplastic tissues. The authors of [8] previously proposed adding a geometrical verification to take spatial information into account, however their method is based on the assumption that they want to retrieve images of the exact same scene, which is not the case for our application. In like manner, our idea is to introduce a geometrical verification process after the retrieval process, but based on the assumption that the spatial relationships between the local features are only statistically the same in the images with similar appearance. To introduce spatial information, we took advantage of the dense property to define the adjacency between two visual words as the 8-adjacency between the two disk regions that are labeled by them on the detection grid. Thus, we are able to store in a co-occurence matrix M of size $K \times K$ the probability for each pair of visual words of being adjacent to each other. In order to best differentiate the images of the benign class from the images of the pathological class, we looked at the most discriminative linear combination W of some elements m of M. This is achieved by a linear discriminant analysis (LDA) which uses the textual diagnostic information in the database. The LDA weights are given by $W = \Sigma^{-1} (\mu_1 - \mu_2)$, where Σ is the covariance matrix of the elements m of M in all training images and μ_i is the mean of the elements m of M in the training image belonging to the class i. From these weights W, we computed the spatial criterion $\alpha = Wm$ for each retrieved image and compared it with the α value of the image of interest. By thresholding the α value during a verification process, outliers are rejected and the first retrieved training images are more relevant.

Expert physicians pointed out that some characteristic global patterns are too partially visible on single still pCLE images to make a robust diagnosis: two still images may have a very similar appearance but be attached to contradictory diagnoses [9]. Fig. 3 shows some bad retrieval results caused by the small-FOV problem. To address this problem, the time dimension of pCLE videos needs to be exploited, by including in the retrieval process the temporal relationship between successive images from the same video sequence. The study reported by [10] proposes a method for video retrieval using the spatial layout of the image regions, but this method has been designed for object matching, which is not our objective. Since successive frames from pCLE videos are only related by viewpoint changes, our approach uses the image mosaicing of [3] to project the temporal dimension of a video sequence onto one image with a larger FOV and of higher resolution. Thus, mosaics can be queried and retrieved in the same way as still images.

4   Experiments and Discussion

At the Mayo Clinic in Jacksonville, the Cellvizio® system, MKT, Paris, was used to image colonic polyps during surveillance colonoscopies in 54 patients. On each acquired video sequence, expert physicians established a pCLE diagnosis [7] that differentiates pathological sequences from benign ones. The video sequences contain from 5 to over a thousand frames and each frame is an image of diameter 500 pixels corresponding to a FOV of 240 μm. To build our pCLE database, we considered a subset of these sequences by discarding those whose quality was insufficient to perform a reliable diagnosis. In each of the remaining 52 video sequences, we selected groups of successive frames according to the length of the sequence. The resulting database is composed of $N = 1036$ still pCLE images and $N' = 66$ pCLE mosaics, half of the data coming from benign sequences and half from pathological ones.

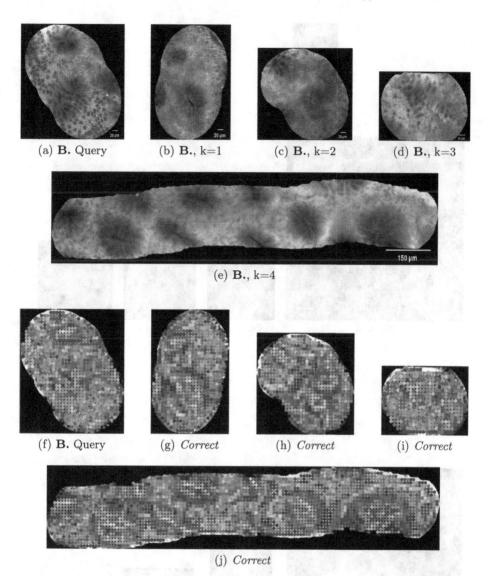

Fig. 4. Typical mosaic retrieval results provided by our method from one benign query. **B.** indicates Benign and **N.** Neoplastic. From left to right on each row: the queried mosaic, and its k-NNs on the top layer, and their respective colored visual words on the bottom layer. FOV of the mosaics: from 260 μm to 1300 μm.

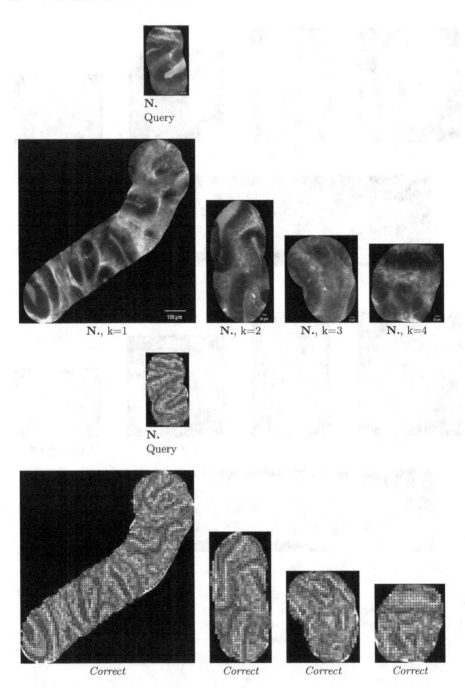

Fig. 5. Typical mosaic retrieval results provided by our method from one neoplastic query. **B.** indicates Benign and **N.** Neoplastic. From left to right on each row: the queried mosaic, and its k-NNs on the top layer, and their respective colored visual words on the bottom layer. FOV of the mosaics: from 260 µm to 1300 µm.

In pCLE images, the disk regions containing information at mesoscopic scale have a radius value $\rho_1 = 40$, while the radius value of those containing information at microscopic scale is $\rho_2 = 15$. For the dense detector, we then chose $\delta = 20$ pixels of grid spacing in order to get a reasonable overlap between adjacent regions. Among the values from 30 to 1500 found in the literature for the number K of visual words provided by the K-Means clustering, the value $K = 100$ yielded satisfying classification results. To prevent overfitting, as the size of our pCLE database is still rather small, especially concerning the number of mosaics, the number of LDA weights in the computation of the spatial criterion α had to be restricted. For the elements m of the co-occurence matrix M, we only considered the K diagonal elements of the matrix M build from the visual words of large radius 40, observing that the overlapping regions of radius 40 have a sufficient spatial correlation, better than those of radius 15. The good values of the threshold θ_α were chosen by analysing the distribution of α across the benign and pathological images: 2 when retrieving still images from a queried still image and 0.5 when retrieving still images from a queried mosaic.

The classification results of our method are presented in Fig. 6 on the left and compared with the following methods taken as references: the standard sparse scale invariant SIFT method, the statistical approach of Haralick features [11,12] and the texture retrieval method of Textons [13]. To ensure a non-biased classification, our validation scheme retrieves k nearest images in the training set with training images not belonging to the video sequence of the image being queried, i.e. a leave-n-out cross-validation where n is the number of frames in the video of the queried image. According to the accuracy, sensitivity and specificity rates yielded by each method on the still images of the pCLE database, our retrieval method including spatial information is the most efficient, with an accuracy rate of 78.2% for $k = 22$ neighbors, which is 11.5 points better than the standard SIFT method. The gain of accuracy can be decomposed in 10.2 points for the choice of a dense detector and a bi-scale SIFT description, and 1.3 points for the verification process on the spatial criterion. It is also worth mentioning that with the spatial verification, fewer nearest neighbors are necessary to classify the query at a given accuracy. For $k = 4$ neighbors, some illustrative examples of the image retrieval results are shown in Fig. 1, 2 and 3.

Moreover, when including both spatial and temporal information by querying mosaics, our classification results are much better, see Fig. 6 on the right. Since mosaics contain more information than single images, their content-based neighborhood is more representative of their pathological neighborhood, so they can be better classified by a smaller number k' of nearest neighbors. Indeed, if we retrieve still images for queried mosaics, the classification accuracy is 83.3% for $k' = 10$ neighbors, which demonstrates the robustness of our retrieval method applied on heterogeneous data with different resolution. For the retrieval of still images from queried mosaics, the poor specificity can be explained by the fact that a mosaic annotated as neoplastic may contain some benign patterns which induce the retrieval of single benign images and classify it as benign. However the expert physicians diagnose a pCLE video sequence as neoplastic as soon as it

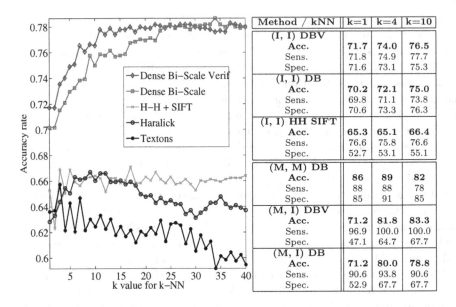

The figure contains a plot on the left and a table on the right.

Method / kNN	k=1	k=4	k=10
(I, I) DBV			
Acc.	**71.7**	**74.0**	**76.5**
Sens.	71.8	74.9	77.7
Spec.	71.6	73.1	75.3
(I, I) DB			
Acc.	**70.2**	**72.1**	**75.0**
Sens.	69.8	71.1	73.8
Spec.	70.6	73.3	76.3
(I, I) HH SIFT			
Acc.	**65.3**	**65.1**	**66.4**
Sens.	76.6	75.8	76.6
Spec.	52.7	53.1	55.1
(M, M) DB			
Acc.	**86**	**89**	82
Sens.	88	88	78
Spec.	85	91	85
(M, M) DBV			
Acc.	**71.2**	**81.8**	**83.3**
Sens.	96.9	100.0	100.0
Spec.	47.1	64.7	67.7
(M, I) DB			
Acc.	**71.2**	**80.0**	**78.8**
Sens.	90.6	93.8	90.6
Spec.	52.9	67.7	67.7

Plot legend: Dense Bi–Scale Verif, Dense Bi–Scale, H–H + SIFT, Haralick, Textons. Y-axis: Accuracy rate. X-axis: k value for k–NN.

Fig. 6. Left: Classification accuracies yielded by several methods on the still images of the pCLE database, with leave-n-out cross-validation. Right: Results for **k** nearest neighbors, where **M** means **Mosaic** and **I** means **Image** in the configuration (**Queried, Retrieved**). Our proposed method is referred to as Dense Bi-Scale Verif (**DBV**) if it includes spatial verification and Dense Bi-Scale (**DB**) otherwise.

contains neoplastic patterns, even when some benign tissue is imaged. Besides, if we retrieve mosaics for queried mosaics, the classification accuracy is 89%. Thus, even though we only have a small number of mosaics, including time dimension in mosaics provides us proof of concept results for endomiscroscopic video retrieval. The relevance of mosaic retrieval can also be qualitatively appreciated in Fig. 4 and 5. For the retrieval of mosaics from queried mosaics, including the spatial information does not improve the classification results because of the overfitting phenomenon: indeed, the number of LDA weights, 100, is bigger than the total number of mosaics, $N' = 66$.

5 Conclusion

Using visual similarity between a given image and medically interpreted images allowed us to provide the physicians with semantic similarity, and thus could potentially support their diagnostic decision. Although our experiments are focused on a relatively small training dataset, the classification results constitute a validation of our generic methodology. By taking into account the spatio-temporal relationship between the local feature descriptors, the first retrieved endomicroscopic images are much more relevant.

For future work, a larger training database would not only improve the classification results if all the characteristics of the image classes are better represented, but also enable the exploitation of the whole co-occurence matrix of visual words at several scales. Besides, the learning step of the retrieval process could leverage the textual information of the database and incorporate the spatial information of multi-scale co-occurence matrices into descriptors. The co-occurence matrix could also be better analyzed by other tools more generic than linear discriminant analysis. For example, a more complete spatial geometry between local features could be learned by estimating the parameters of Markov Random Fields [14]. We also plan to perform a leave-one-patient-out cross-validation to ensure fully unbiased retrieval and classification results. As for introducing the temporal information, a more robust approach would not only consider the fused image of a mosaic but the $2D + t$ volume of the registered frames composing the mosaic to work on more accurate visual words and better combine spatial and temporal information.

References

1. Wallace, M., Fockens, P.: Probe-based confocal laser endomicroscopy. Gastroenterology 136(5), 1509–1513 (2009)
2. Zhang, J., Lazebnik, S., Schmid, C.: Local features and kernels for classification of texture and object categories: a comprehensive study. International Journal of Computer Vision 73, 213–238 (2007)
3. Vercauteren, T., Perchant, A., Malandain, G., Pennec, X., Ayache, N.: Robust mosaicing with correction of motion distortions and tissue deformation for in vivo fibered microscopy. Medical Image Analysis 10(5), 673–692 (2006)
4. Mikolajczyk, K., Tuytelaars, T., Schmid, C., Zisserman, A., Matas, J., Schaffalitzky, F., Kadir, T., Van Gool, L.: A comparison of affine region detectors. International Journal of Computer Vision 65, 43–72 (2005)
5. Lowe, D.: Distinctive image features from scale-invariant keypoints. International Journal of Computer Vision 60, 91–110 (2004)
6. André, B., Vercauteren, T., Perchant, A., Wallace, M.B., Buchner, A.M., Ayache, N.: Endomicroscopic image retrieval and classification using invariant visual features. In: Proceedings of the IEEE International Symposium on Biomedical Imaging: From Nano to Macro (ISBI 2009), pp. 346–349 (2009)
7. Buchner, A., Ghabril, M., Krishna, M., Wolfsen, H., Wallace, M.: High-resolution confocal endomicroscopy probe system for in vivo diagnosis of colorectal neoplasia. Gastroenterology 135(1), 295 (2008)
8. Jegou, H., Douze, M., Schmid, C.: Hamming embedding and weak geometric consistency for large scale image search. In: Forsyth, D., Torr, P., Zisserman, A. (eds.) ECCV 2008, Part I. LNCS, vol. 5302, pp. 304–317. Springer, Heidelberg (2008)
9. Becker, V., Vercauteren, T., von Weyern, C.H., Prinz, C., Schmid, R.M., Meining, A.: High resolution miniprobe-based confocal microscopy in combination with video-mosaicing. Gastrointestinal Endoscopy 66(5), 1001–1007 (2007)
10. Sivic, J., Zisserman, A.: Efficient visual search for objects in videos. Proceedings of the IEEE 96, 548–566 (2008)
11. Haralick, R.: Statistical and structural approaches to texture. Proceedings of the IEEE 67, 786–804 (1979)

12. Srivastava, S., Rodriguez, J., Rouse, A., Brewer, M., Gmitro, A.: Computer-aided identification of ovarian cancer in confocal microendoscope images. Journal of Biomedical Optics 13(2), 024021 (2008)
13. Leung, T., Malik, J.: Representing and recognizing the visual appearance of materials using three-dimensional textons. International Journal of Computer Vision 43, 29–44 (2001)
14. Descombes, X., Morris, R., Zerubia, J., Berthod, M.: Estimation of Markov random field prior parameters using Markov chain Monte Carlo maximum likelihood. IEEE Transactions on Image Processing 8(7), 954–963 (1999)

A Query-by-Example Content-Based Image Retrieval System of Non-melanoma Skin Lesions

Lucia Ballerini[1], Xiang Li[1], Robert B. Fisher[1], and Jonathan Rees[2]

[1] School of Informatics, University of Edinburgh, UK
x.li-29@sms.ed.ac.uk, lucia.ballerini@ed.ac.uk, rbf@inf.ed.ac.uk
[2] Dermatology, University of Edinburgh, UK
jonathan.rees@ed.ac.uk

Abstract. This paper proposes a content-based image retrieval system for skin lesion images as a diagnostic aid. The aim is to support decision making by retrieving and displaying relevant past cases visually similar to the one under examination. Skin lesions of five common classes, including two non-melanoma cancer types are used. Colour and texture features are extracted from lesions. Feature selection is achieved by optimising a similarity matching function. Experiments on our database of 208 images are performed and results evaluated.

1 Introduction

Research in content-based image retrieval (CBIR) today is an extremely active discipline. There are already review articles containing references to a large number of systems and description of the technology implemented [1,2]. A more recent review [3] reports a tremendous growth in publications on this topic. Applications of CBIR systems to medical domains already exist [4], although most of the systems currently available are based on radiological images.

Most of the work in dermatology has focused on skin cancer detection. Different techniques for segmentation, feature extraction and classification have been reported by several authors. Concerning segmentation, Celebi et al. [5] presented a systematic overview of recent border detection methods: clustering followed by active contours are the most popular. Numerous features have been extracted from skin images, including shape, colour, texture and border properties [6,7,8]. Classification methods range from discriminant analysis to neural networks and support vector machines [9,10,11]. These methods are mainly developed for images acquired by epiluminescence microscopy (ELM or dermoscopy) and they focus on melanoma, which is actually a rather rare, but quite dangerous, condition whereas other skin cancers are much more common.

To our knowledge, there are few CBIR systems in dermatology. Chung et al. [12] created a skin cancer database. Users can query the database by feature attribute values (shape and texture), or by synthesised image colours. It does not include a query-by-example method, as do most common CBIR systems. The report concentrates on the description of the web-based browsing and data

B. Caputo et al. (Eds.): MCBR-CDS 2009, LNCS 5853, pp. 31–38, 2010.

mining. However, nothing is said about database details (number, lesion types, acquisition technique), nor about the performance of the retrieval system. Celebi et al. [13] developed a system for retrieving skin lesion images based on shape similarity. The novelty of that system is the incorporation of human perception models in the similarity function. Results on 184 skin lesion images show significant agreement between computer assessment and human perception. However, they only focus on silhouette shape similarity and do not include many features (colour and texture) described in other papers by the same authors [11]. Rahman et al. [14] presented a CBIR system for dermatoscopic images. Their approach include image processing, segmentation, feature extraction (colour and textures) and similarity matching. Experiments on 358 images of pigmented skin lesions from three categories (benign, dysplastic nevi and melanoma) are performed. A quantitative evaluation based on the precision curve shows the effectiveness of their system to retrieve visually similar lesions (average precision $\simeq$ 60%). Dorileo et al. [15] presented a CBIR system for wound images (necrotic tissue, fibrin, granulation and mixed tissue). Features based on histogram and multispectral co-occurrence matrices are used to retrieve similar images. The performance is evaluated based on measurements of precision ($\simeq$ 50%) on a database of 215 images. All these approaches only consider a few classes of lesions and/or do not exploit many useful features in this context.

Dermatology atlases containing a large number of images are available on line [16,17]. However, their searching tool only allows query by the name of the lesion. On the other hand, the possibility of retrieving images based on visual similarity would greatly benefit both the non-expert users and the dermatologists. As already pointed out [4,14], there is a need for CBIR as a decision support tool for dermatologists in the form of a display of relevant past cases, along with proven pathology and other suitable information. CBIR could be used to present cases that are not only similar in diagnosis, but also similar in appearance and cases with visual similarity but different diagnoses. Hence, it would be useful as a training tool for medical students and researchers to browse and search large collection of disease related illustrations using their visual attributes.

Motivated by this, we propose a CBIR approach for skin lesion images. The present work focuses on 5 common classes of skin lesions: Actinic Keratosis (AK), Basal Cell Carcinoma (BCC), Melanocytic Nevus / Mole (ML), Squamous Cell Carcinoma (SCC), Seborrhoeic Keratosis (SK). As far we know, **this is the first query-by-example CBIR system for these 5 classes of lesions**. Our system mainly relies on colour and texture features, and gives values of precision between 59% and 63%.

2   Feature Extraction

CBIR undertakes the extraction of several features from each image, which, consequently, are used for computing similarity between images during the retrieval procedure. These features describe the content of the image and that is why they must be appropriately selected according to the context. The features have to be discriminative and sufficient for the description of different pathologies. Basically, the key to attain a successful retrieval system is to choose the right features that represent each class of images as uniquely as possible.

Many feature selection and extraction strategies have been proposed [6,7] from the perspective of classification of images as malignant or benign. Different features attempt to reflect the parameters used in medical diagnosis, such as $ABCD$ rule for melanoma detection [18]. These features are certainly effective for the classification purpose, as seen from the performance of some classification-based systems in this domain, claiming a correct classification up to 100% [10] or specificity/sensitivity of 92.34%/93.33% [11]. However, features good for classification or distinguishing one disease from another may not be suitable for retrieval and display of similar appearing lesions. In this retrieval system, we are looking for similar images in term of colour, texture, shape, etc. By selecting and extracting good representative features, we may be able to identify images similar to an unknown query image, whether it belongs to the same disease group or not. Similar images belonging to different classes may give an idea about the certainty of classification.

Skin lesions appear mainly characterised by their colour and texture. In the rest of this section we will describe the features that can describe such properties, as well as some normalisation procedures employed to make sure only the lesion colour information is taken into consideration.

Colour. Colour features are represented by the mean colour $\mu = (\mu_R, \mu_G, \mu_B)$ of the lesion and their covariance matrix Σ. Let

$$C_{XY} = \frac{1}{N} \left[\sum_{i=1}^{N} X_i Y_i \right] - \mu_X \mu_Y \tag{1}$$

where: N is the number of pixels in the lesion, X_i the colour component of channel X ($X, Y \in \{R, G, B\}$) of pixel i. Assuming to use the original RGB (Red, Green, Blue) colour space, the covariance matrix is:

$$\Sigma = \begin{bmatrix} C_{RR} & C_{RG} & C_{RB} \\ C_{GR} & C_{GG} & C_{GB} \\ C_{BR} & C_{BG} & C_{BB} \end{bmatrix} \tag{2}$$

In this work, RGB, HSV (Hue, Saturation, Value) and CIE_Lab, CIE_Lch (Munsell colour coordinate system [14]) and Otha [19] colour spaces are used.

A number of normalisation techniques have been applied before extracting colour features. We normalised each colour component by the average of the same component of the safe skin of the same patient, because it had best performance.

After experimenting with the 5 different colour spaces, we selected the normalised RGB, because it was slightly better, but there was not a huge difference.

Texture. Texture features are extracted from generalised co-occurrence matrices. Assume an image I having N_x columns, N_y rows and N_g grey levels.

Let $L_x = \{1, 2, \cdots, N_x\}$ be the columns, $L_y = \{1, 2, \cdots, N_y\}$ be the rows, and $G_x = \{0, 1, \cdots, N_g - 1\}$ be the set of quantised grey levels. The co-occurrence matrix P_δ is a matrix of dimension $N_g \times N_g$, where [20]:

$$P_\delta(i, j) = \#\{((k, l), (m, n)) \in (L_y \times L_x) \times (L_y \times L_x) | I(k, l) = i, I(m, n) = j\} \quad (3)$$

i.e. the number of co-occurrences of the pair of grey level i and j which are a distance $\delta = (d, \theta)$ apart. In our work, the pixel pairs (k, l) and (m, n) have distance $d = 1, \cdots, 6$ and orientation $\theta = 0°, 45°, 90°, 135°$.

Generalised co-occurrence matrices (GCM) are the extension of the co-occurrence matrix to multispectral images, i.e. images coded on n colour channels. Let u and v be two colour channels. The generalised co-occurrence matrices are:

$$P_\delta^{(u,v)}(i, j) = \#\{((k, l), (m, n)) \in (L_y \times L_x) \times (L_y \times L_x) | I_u(k, l) = i, I_v(m, n) = j\} \quad (4)$$

For example, in case of colour image, coded on three channels (RGB), we have six cooccurrence matrices: (RR),(GG),(BB) that are the same as grey level co-occurrence matrices computed on one channel and (RG), (RB), (GB) that take into account the correlations between the channels.

In order to have orientation invariance for our set of GCMs, we averaged the matrices with respect to θ. Quantisation levels $N_G = 64, 128, 256$ are used for the three colour spaces: RGB, HSV and CIE_Lab.

From each GCM we extracted 12 texture features: energy, contrast, correlation, entropy, homogeneity, inverse difference moment, cluster shade, cluster prominence, max probability, autocorrelation, dissimilarity and variance as defined in [20], for a total of 3888 texture features (12 features × 6 inter-pixel distances × 6 colour pairs × 3 colour spaces × 3 grey level quantisations).

Texture features are also extracted from the sum- and difference-histograms (SDHs) as proposed by Unser [21]. These histograms are defined as:

$$h_S(i) = \#\{((k, l), (m, n)) \in (L_y \times L_x) \times (L_y \times L_x) | I(k, l) + I(m, n) = i\} \quad (5)$$
$$h_D(j) = \#\{((k, l), (m, n)) \in (L_y \times L_x) \times (L_y \times L_x) | I(k, l) - I(m, n) = j\} \quad (6)$$

As for the co-occurrence matrices, we generalised the SDHs by considering the intra- and inter-plane sum- and difference-histograms:

$$h_{S,D}^{(u,v)}(i) = \#\{((k, l), (m, n)) \in (L_y \times L_x) \times (L_y \times L_x) | I_u(k, l) \pm I_v(m, n) = i\} \quad (7)$$

Similarly to the GCMs, we constructed a set of SDHs varying pixel displacement, orientation, quantisation level, and colour spaces.

From each SDH we extracted 15 features: sum mean, sum variance, sum energy, sum entropy, diff mean, diff variance, diff energy, diff entropy, cluster shade, cluster prominence, contrast, homogeneity, correlation, angular second moment, entropy as defined in [21], as well as the relative illumination invariant features described by Münzenmayer [22], for a total of other 9720 features (15 features × 2 illumination invariants × 6 inter-pixel distances × 6 colour pairs × 3 colour spaces × 3 grey level quantisations).

3 Similarity Matching

The retrieval system is based on a similarity measure defined between the query image Q and a database image I.

For colour covariance-based features, the Bhattacharyya distance metric is used as follow:

$$D_C(Q,I) = \frac{1}{8}(\mu_Q - \mu_I)^T \left[\frac{(\Sigma_Q + \Sigma_I)}{2} \right]^{-1} (\mu_Q - \mu_I) + \frac{1}{2} \ln \frac{\left| \frac{(\Sigma_Q + \Sigma_I)}{2} \right|}{\sqrt{|\Sigma_Q||\Sigma_I|}} \qquad (8)$$

where μ_Q and μ_I are the average colour feature vectors, Σ_Q and Σ_I are the covariance matrices of the lesion of Q and I respectively, and $|\cdot|$ denotes the matrix determinant.

The Euclidean distance $D_T(Q,I)$ is used for distances between a subset of texture features f_{subset}, selected as described later.

$$D_T(Q,I) = \|f_{subset}^Q - f_{subset}^I\| = \sqrt{\sum_{i=1}^{m}(f_i^Q - f_i^I)^2} \qquad (9)$$

Other metric distances have been considered, but gave worse results.

We aggregated the two distances into a similarity matching function as:

$$S(Q,I) = w_C \cdot D_C(Q,I) + (1 - w_C) \cdot D_T(Q,I) \qquad (10)$$

where w_C and w_T are weighting factors that need to be selected experimentally. In our case, $w_C = 0.7$ gave the best results.

3.1 Feature Selection

Feature selection is applied in order to select a subset of texture features from all the feature extracted. Two feature selection algorithms have been implemented: a greedy algorithm and a genetic algorithm (GA) [23]. In both cases, the objective function is the maximisation of the number of correctly retrieved images, i.e. the images belonging to the same class as the query image. This measure is closely related to *precision*, that is the ratio of the number of relevant images returned to the total number of images returned. We averaged it using each image in the database as query image, and asking the system to retrieve 10 similar images for each presented image (not retrieving itself). Often, in the information retrieval context, the F-measure, that is a weighted harmonic mean of precision and recall is often used. In our case, due to the uneven distribution of lesions into classes, it seemed more appropriate to maximise the precision.

The greedy algorithm was stopped once 10 features were selected. The greedy algorithm was slightly modified so that features with high correlation with any of the already selected features were eliminated.

In the genetic algorithm, the feature indexes are encoded in the chromosomes as integer numbers. Each chromosome contains 10 features. Other GA parameters (determined after a number of experiments varying such parameters) are: 1000 individuals, 0.9 crossover rate, adaptive feasible mutation, 100 generations, 10 runs. The results reported below are based on the GA as it had higher performance.

4   Results and Evaluation

The effectiveness of the proposed retrieval system is evaluated on our image database of 208 lesions, belonging to 5 classes (11 AK, 59 BCC, 58 ML, 18 SCC, 62 SK). Images are acquired using a Canon EOS 350D SRL camera, having a resolution of about 0.03 mm. Lesions are manually segmented in this study. Ground truth is provided by the medical co-authors.

Typical screen-shots of our CBIR system are shown in Figure 1.

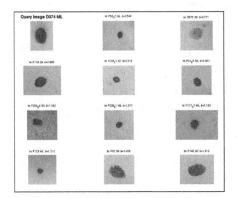 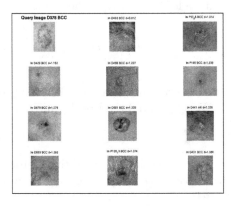

Fig. 1. Two screenshots showing retrieved images similar to the query image (top left image in each screenshot). Respectively 5/11 and 10/11 are correctly retrieved.

For medical image retrieval systems, the evaluation issue is very often neglected in most of the papers [4]. We show average precision/recall and precision/scope curves obtained by evaluating top N retrieved results (scope). These graphs justify our choices of best parameters and techniques for our system.

Figure 2(a) shows the precision/recall curves obtained using colour and texture features, where a subset of the latter ones are selected by GA (which outperforms the greedy feature selection). Note that colour and texture feature are co-optimised so there may be some better independent colour and texture feature sets.

Figure 2(b) shows the precision/scope curves using our method, and the average curve obtained by a similar retrieval system [14]. We estimated the average curve assuming even lesion distribution between the three classes they consider. Our system precision is about 8% higher of their system one. However a fair comparison is not possible, because their lesion classes are completely different from ours, only one of which (mole) is one of the classes we consider.

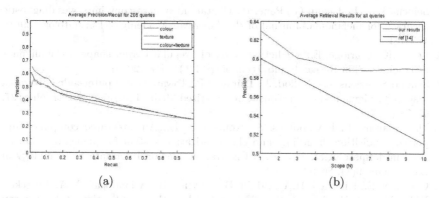

Fig. 2. (a) Precision/Recall curves using colour and texture features, (b) Precision/Scope curves using our method and the one reported in ref. [14]

5   Conclusions

We have presented a CBIR system as a diagnostic aid for skin lesion images. We believe that presenting images with known pathology that are visually similar to an image being evaluated may provide intuitive clinical decision support to dermatologists. Further studies will include the extraction of other texture-related features (i.e. fractal dimension, Gabor- and Tamura-based) as well as shape and boundary features. A larger database is being reached. We plan also to include relevance feedback, which is commonly used in image retrieval, but has not yet been used for medical images.

Acknowledgements. We thank the Wellcome Trust for funding this project.

References

1. Rui, Y., Huang, T.S., Chang, S.F.: Image retrieval: Current techniques, prominsign directions, and open issues. Journal of Visual Communication and Image Representation 10, 39–62 (1999)
2. Smeulders, A.W.M., Member, S., Worring, M., Santini, S., Gupta, A., Jain, R.: Content-based image retrieval at the end of the early years. IEEE Transactions on Pattern Analysis and Machine Intelligence 22(12), 1349–1380 (2000)
3. Datta, R., Joshi, D., Li, J., Wang, J.Z.: Image retrieval: Ideas, influences, and trends of the new age. ACM Computing Surveys 40(2), 1–5 (2008)
4. Müller, H., Michoux, N., Bandon, D., Geissbuhler, A.: A review of content-based image retrieval systems in medical applications - clinical benefits and future directions. International Journal of Medical Informatics 73, 1–23 (2004)
5. Celebi, M.E., Iyatomi, H., Schaefer, G., Stoecker, W.V.: Lesion border detection in dermoscopy images. Computerized Medical Imaging and Graphics 33(2), 148–153 (2009)
6. Wollina, U., Burroni, M., Torricelli, R., Gilardi, S., Dell'Eva, G., Helm, C., Bardey, W.: Digital dermoscopy in clinical practise: a three-centre analysis. Skin Research and Technology 13, 133–142 (2007)

7. Seidenari, S., Pellacani, G., Pepe, P.: Digital videomicroscopy improves diagnostic accuracy for melanoma. Journal of the American Academy of Dermatology 39(2), 175–181 (1998)
8. Lee, T.K., Claridge, E.: Predictive power of irregular border shapes for malignant melanomas. Skin Research and Technology 11(1), 1–8 (2005)
9. Schmid-Saugeons, P., Guillod, J., Thiran, J.P.: Towards a computer-aided diagnosis system for pigmented skin lesions. Computerized Medical Imaging and Graphics 27, 65–78 (2003)
10. Maglogiannis, I., Pavlopoulos, S., Koutsouris, D.: An integrated computer supported acquisition, handling, and characterization system for pigmented skin lesions in dermatological images. IEEE Transactions on Information Technology in Biomedicine 9(1), 86–98 (2005)
11. Celebi, M.E., Kingravi, H.A., Uddin, B., Iyatomi, H., Aslandogan, Y.A., Stoecker, W.V., Moss, R.H.: A methodological approach to the classification of dermoscopy images. Computerized Medical Imaging and Graphics 31(6), 362 (2007)
12. Chung, S.M., Wang, Q.: Content-based retrieval and data mining of a skin cancer image database. In: International Conference on Information Technology: Coding and Computing (ITCC 2001), pp. 611–615. IEEE Computer Society, Los Alamitos (2001)
13. Celebi, M.E., Aslandogan, Y.A.: Content-based image retrieval incorporating models of human perception. In: International Conference on Information Technology: Coding and Computing, vol. 2, p. 241 (2004)
14. Rahman, M.M., Desai, B.C., Bhattacharya, P.: Image retrieval-based decision support system for dermatoscopic images. In: IEEE Symposium on Computer-Based Medical Systems, pp. 285–290. IEEE Computer Society, Los Alamitos (2006)
15. Dorileo, E.A.G., Frade, M.A.C., Roselino, A.M.F., Rangayyan, R.M., Azevedo-Marques, P.M.: Color image processing and content-based image retrieval techniques for the analysis of dermatological lesions. In: 30th Annual International Conference of the IEEE Engineering in Medicine and Biology Society (EMBS 2008), August 2008, pp. 1230–1233 (2008)
16. Dermnet: the dermatologist's image resource, Dermatology Image Altas (2007), http://www.dermnet.com/
17. Cohen, B.A., Lehmann, C.U.: Dermatlas (2000-2009) Dermatology Image Altas, http://dermatlas.med.jhmi.edu/derm/
18. Johr, R.H.: Dermoscopy: alternative melanocytic algorithms–the abcd rule of dermatoscopy, menzies scoring method, and 7-point checklist. Clinics in Dermatology 20(3), 240–247 (2002)
19. Ohta, Y.I., Kanade, T., Sakai, T.: Color information for region segmentation. Computer Graphics and Image Processing 13(1), 222–241 (1980)
20. Haralick, R.M., Shanmungam, K., Dinstein, I.: Textural features for image classification. IEEE Transactions on Systems, Man and Cybernetics 3(6), 610–621 (1973)
21. Unser, M.: Sum and difference histograms for texture classification. IEEE Transactions on Pattern Analysis and Machine Intelligence 8(1), 118–125 (1986)
22. Munzenmayer, C., Wilharm, S., Hornegger, J., Wittenberg, T.: Illumination invariant color texture analysis based on sum- and difference-histograms. In: Kropatsch, W.G., Sablatnig, R., Hanbury, A. (eds.) DAGM 2005. LNCS, vol. 3663, pp. 17–24. Springer, Heidelberg (2005)
23. Goldberg, D.E.: Genetic Algorithms in Search, Optimization, and Machine Learning. Addison-Wesley, Reading (1989)

3D Case–Based Retrieval for Interstitial Lung Diseases

Adrien Depeursinge[1], Alejandro Vargas[1], Alexandra Platon[2],
Antoine Geissbuhler[1], Pierre–Alexandre Poletti[2], and Henning Müller[1,3]

[1] Medical Informatics, Geneva University Hospitals and University of Geneva, CH
[2] Service of Emergency Radiology, University Hospitals of Geneva, CH
[3] Business Information Systems, University of Applied Sciences Sierre, CH

Abstract. In this paper, a computer–aided diagnosis (CAD) system that
retrieves similar cases affected with an interstitial lung disease (ILDs) to
assist the radiologist in the diagnosis workup is presented and evaluated.
The multimodal inter–case distance measure is based on a set of clinical
parameters as well as automatically segmented 3–dimensional regions of
lung tissue in high–resolution computed tomography (HRCT) of the chest.
A global accuracy of 75.1% of correct matching among five classes of lung
tissues as well as a mean average retrieval precision at rank 1 of 71% show
that automated lung tissue categorization in HRCT data is complementary to case–based retrieval both from the user's viewpoint and also on the
algorithmic side.

1 Introduction

Interstitial lung diseases (ILDs) can be characterized by the gradual alteration of
the lung parenchyma leading to breathing dysfunction. They regroup more than
150 disorders of the lung parenchyma. The factors and mechanisms of the disease
processes vary from one disease to another. The diagnosis of these pathologies
is established based on the complete history of the patient, a physical examination, laboratory tests, pulmonary function testing as well as visual findings on
chest X–ray. When the synthesis of this information arouses suspicions toward an
ILD, high–resolution computer tomography (HRCT) imaging of the chest is often required to acquire a rapid and accurate visual assessment of the lung tissue.
Compared to the radiograph, the tomographic image acquisition process avoids
the superposition of the lung tissue with the ribs and other organs leading to
three–dimensional images of the lung volumes. Nevertheless, the interpretation
of HRCT is often challenging with numerous differential diagnoses and it is currently reserved to experienced radiologists [1]. A correct interpretation requires
an advanced knowledge of the lung anatomy and the alterations of the interstitial tissues have wide intra and inter–disease variations. Moreover, the large
number of slices contained in an HRCT image series makes the interpretation
task time–consuming and subject to missing relevant lung tissue patterns.

Recent advances in medical informatics enabled access to most of the radiological exams to all clinicians through the electronic health record (EHR) and the

B. Caputo et al. (Eds.): MCBR-CDS 2009, LNCS 5853, pp. 39–48, 2010.
© Springer-Verlag Berlin Heidelberg 2010

picture archival and communication system (PACS). This change of the medical workflow calls upon computer expert systems able to bring the right information to the right people at the right time. Interpreting HRCT of the chest is no exception to this and has been an active research domain during the last decade [2,3,4,5]. Most of the proposed systems aim at categorizing lung tissue to provide a second opinion to radiologists. This provides a quick and exhaustive scan of the large number of images and can draw the radiologist's attention on diagnostically useful parts of the images. To be useful in clinical practice such systems have to be able to detect a sufficient number of types of lung tissue [6]. Several studies obtained recognition rates of up to 90% correct matches [7] but usually while training and testing the classifier with images belonging to the same patients, which introduces a large bias compared to a real–world clinical usage with unknown incoming images [8].

In the context of medical image analysis, providing quick and precious information to the clinician is not limited to automatic recognition of abnormal tissue and/or structure. The approach of the clinician to a diagnosis if he has little experience of the domain is to compare the image under investigation with typical cases with confirmed diagnosis listed in textbooks or contained in personal collections. It allows to rule out diagnostics and, in association with clinical parameters, prevents the reader from mixing diagnoses with similar radiological findings. This process allows the clinician to partly replace a lack of experience but has two major drawbacks: searching for similar images is time–consuming and the notion of similarity may be subjective and can be ambiguous [9].

Content–based medical image retrieval (CBIR) aims at finding objectively visually similar images in large standardized image collections such as PACS [10]. For instance, the Radiology Department of the University Hospitals of Geneva (HUG) alone produced more than 80,000 images a day in 2008, representing retrospectively a potentially large repository of knowledge and experience as images are all associated with one or several diagnosis. The notion of similarity is usually established from a set of visual features describing the content of the images. Features can vary from low–level measures such as the histogram quantification of the colors to high–level semantically–related features describing the anatomical content of images. Few CBIR systems have been evaluated in clinical practice but some of them showed that they can be accepted by the clinicians as a useful tool [11,12]. The use of a CBIR system clearly increased the number of correct diagnoses within the context of the interpretation of HRCT images associated with ILDs in [13]. A possible extension to CBIR is to carry out case–based retrieval. Most often, the clinician actually looks for similar cases as he considers the image within the context of a patient with a personal history, findings on the physical examination, laboratory tests, etc. Radiologists never interpret an image without taking into account the clinical context defined by disease–specific metadata.

In this paper we show that automated lung classification in HRCT data is complementary to case–based retrieval, both from the user's viewpoint and also on the algorithmic side. In a first step, healthy and four abnormal tissue types associated with 7 of the most common ILDs are automatically detected in HRCT. These latter are *emphysema, ground glass, fibrosis, micronodules* and *healthy*. Then, based on the volumes of the segmented tissues and a set of selected clinical parameters, similar cases are retrieved from a multimedia database of ILD cases built at the HUG within the context of the Talisman project[1].

2   Methods

This section describes the various steps of our computer–aided diagnosis (CAD) system for ILDs consisting of semi–automatic segmentation of the lung volumes, classification of the lung tissue based on texture properties, and multimodal retrieval of similar cases.

2.1   Semi–automatic Segmentation of the Lung Volumes

Segmentation of the lung volumes is a required preliminary step to lung tissue categorization. The result of this step is a binary mask M_{lung} that indicates the regions to be analyzed by the texture analysis routines. Since the geometries and shapes of the lungs are subject to large variations among the cases, semi–automatic segmentation based on region growing and mathematical morphology is carried out. The region growing routine contained in YaDiV[2] is used. Starting from a seed point $s(x, y, z)$ defined by the user, each 26–connected neighbor is added to the region M_{lung} if the summed value of its own neighbors differs of less than a given variance v defined by the user. At this stage, M_{lung} describes the global lung regions well but contains many holes where the region growing algorithm was stopped by denser regions such as vessels or consolidations of the lung parenchyma. To fill these holes, a closing operation is applied to M_{lung} using a spherical structuring element. Two parameters require attention from the user: the radius r of the structuring element in millimeters and N_{op} which defines the number of closing operations.

2.2   Automated Lung Tissue Categorization

Our approach for categorizing lung tissue patterns associated with ILDs in HRCT relies on texture analysis. Most of the patterns depict nonfigurative and cellularly organized areas of pixels. To describe texture properties, features based on grey–level histogram in Hounsfield Units (HU) as well as statistical measures from a tailored wavelet transform are extracted. A support vector machine (SVM) classifier is used to draw boundaries among the distinct classes of lung tissue represented in the feature space.

---

[1] Talisman: Texture Analysis of Lung ImageS for Medical diagnostic AssistaNce, http://www.sim.hcuge.ch/medgift/01_Talisman_EN.htm, as of 8 November 2009.

[2] YaDiV: Yet Another DIcom Viewer, http://www.welfenlab.de/en/research/fields_of_research/yadiv/, as of 8 November 2009.

Grey–Level Histogram. CT scanners deliver DICOM images with pixel values in HU that can be univoquely mapped to the density of the observed tissue. Thus, essential information is contained directly in the grey–levels. To encode this information, 22 histogram bins bin_j of grey–levels in $[-1050; 600[$ corresponding to the interval of the lung HU values (including pathological) are used as texture features. An additional feature pix_{air} counts the number of air pixels which have value below -1000 HU.

Wavelet–Based Features. To be complementary to the grey–level histogram, attributes describing the spatial distribution of the pixels are required. Multi-scale analysis using wavelet transforms proved to be adequate for texture analy-sis [14] but requires to control the three essential affine parameters: translation, scale–progression and directionality. For lung tissue analysis, we assume that lung tissues patterns in transversal slices are translation and rotation–invariant. Moreover, a fine and initializable scale–progression is necessary to distinguish between vessels and micronodules. To assess translation invariance, a wavelet frame transform is used. Isotropic polyharmonic B–spline wavelets along with the nonseparable quincunx subsampling scheme yield a near isotropic wavelet transform with fine and tunable scale–progression [15]. The classical separable wavelet transform tends to favor the vertical and horizontal directions, and pro-duces a so–called "diagonal" wavelet component, which does not have a straight-forward directional interpretation. The quincunx scale–progression is finer than the widely used dyadic one with a subsampling factor of $\sqrt{2}$ instead of 2. In addition to be near isotropic by implementing a multiscale smoothed version of the Laplacian Δ, isotropic polyharmonic B–spline wavelets can be scaled easily using the order γ that iterates Δ:

$$\psi_\gamma(\mathbf{D}^{-1}\mathbf{x}) = \Delta^{\frac{\gamma}{2}}\{\phi\}(\mathbf{x}), \tag{1}$$

, where ϕ is an appropriate smoothing (low–pass) function and $\mathbf{D} = [1\,1; 1\,-1]$ is the quincunx subsampling matrix. Statistical measures of the wavelet coefficients are extracted as texture features. Two variances $\sigma_{1,2}^i$ and fixed means $\mu_{1,2}^i = \mu^i$ of a mixture of two Gaussians are estimated using the expectation–maximization (EM) algorithm for each subband $i \in [1; 8]$. The low–pass filtered images are left aside. $\gamma = 3$ obtained the best accuracy of the lung tissue patterns in [16].

Blockwise Classification. In order to automatically categorize every pixel of M_{lung}, each 2D slice is divided into overlapping blocks. Preliminary results using block sizes of $\{8\times8; 16\times16; 24\times24; 32\times32; 40\times40; 48\times48; 56\times56; 64\times64\}$ showed that optimal blocks of size 32×32 is the best trade–off between classification performance and localization. For each block, 22 histogram bins bin_j of GLH in $[-1050; 600]$ and the number of air pixels pix_{air} are concatenated into one hybrid feature vector $\mathbf{v}$ along with GMM parameters of 8 iterations of the quincunx wavelet transform using β_γ of order $\gamma = 3$:

$$\mathbf{v} = (bin_1 \ldots bin_{22} \quad pix_{air} \quad \mu^1\,\sigma_1^1\,\sigma_2^1 \ldots \mu^8\,\sigma_1^8\,\sigma_2^8) \tag{2}$$

An SVM classifier learns from the space spanned by v to find the decision boundaries among five classes of lung tissue. The optimal cost of the errors C and the width of the Gaussian kernel σ_{kernel} are found using a grid search with $C \in [0; 100]$ and $\sigma_{kernel} \in [10^{-2}; 10^2]$.

2.3   Multimodal Case–Based Retrieval

In order to retrieve similar cases from a database to assist the clinician in diagnosis of ILDs, a distance measure based on the volumes of segmented tissue groups as well as on clinical parameters is used. Case–based retrieval is enabled by the automated categorization of the entire HRCT image series. The three–dimensional map of the lung tissue obtained with the blockwise classification of the lung regions yields a semantically–related basis for the comparison of the cases. The percentages of the respective volumes v_i of the five classes of lung tissue are used to assess the visual similarity between HRCT image series from two patients. The respective volumes of lung tissue are semantically related to the ILDs as each histological diagnosis is associated to a given combination of HRCT findings. This allows to reduce the semantic gap between the user's intentions and the visual features, which is often a bottleneck in CBIR [17]. The Euclidean distance is computed from the percentages of the five volumes of tissue as follows:

$$d_{vol} = \sqrt{v_h^2 + v_e^2 + v_g^2 + v_f^2 + v_m^2} \tag{3}$$

with v_h corresponding to *healthy* tissue, v_e to *emphysema*, v_g to *ground glass*, v_f to *fibrosis* to v_m for *micronodules*.

44 clinical parameters with two levels of importance are used to assess the "meta–similarity" between the cases. The levels of importance are defined by a physician according to the relevance for establishing the diagnosis of eight common ILDs. 3 clinical parameters of first importance include age, gender and smoking history. Another 41 parameters of second importance included physical findings, medical history, and laboratory results. The parameters associated with biopsy outcomes were not included as the goal of the CAD is to provide quick information to the radiologists before any biopsy.

The multimodal distance measure d_M between two cases is computed as a linear combination of three modalities:

$$d_M = a_1 d_{vol} + a_2 d_{param1} + a_3 d_{param2}, \tag{4}$$

with a_j being the weights of each modality. d_{vol} is the Euclidean distance in terms of percentages of the volumes of segmented tissue according to (3) and $d_{param1,2}$ the euclidean distance in terms of clinical parameters of importance 1 and 2, respectively. d_{vol} and $d_{param1,2}$ are normalized before being combined in (4).

3   Results

In this section, the techniques described in Section 2 are applied to a multimedia dataset consisting of 76 cases with at least one annotated HRCT image series with

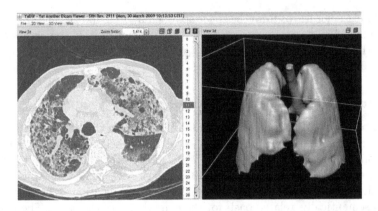

Fig. 1. An example of the segmentation of the lung volumes using a modified version of YaDiV

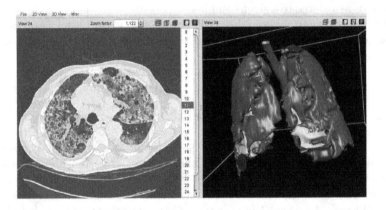

Fig. 2. Automated segmentation of the lung tissue patterns of a patient affected with pulmonary fibrosis. The 3D segmented regions are displayed to the clinician using YaDiV. Green: *healthy* (0.1 liters), blue: *emphysema* (0.39 l), yellow: *ground glass* (0.53 l), red: *fibrosis* (1.91 l), pink: *micronodules* (1.77 l).

slice thickness of $1mm$ and no contrast agent. Annotation of regions was performed together by two experienced radiologists at the HUG. The diagnoses of each of the cases was confirmed either by pathology (biopsy, bronchoalveolar washing) or by a laboratory/specific test confirming the diagnosis. For 69 of the 76 cases, 99 clinical parameters were collected from the EHR, describing the patient's clinical state at the time of the stay when the HRCT image series was acquired. 46 of these parameters were used for the retrieval of similar cases in Equation (4).

To obtain recognition rates of the lung tissue patterns that are representative for actual clinical situations, a leave–one–patient–out cross–validation [8] is used to avoid training and testing the SVM classifier with images belonging to the same patient. All images from the selected case are left aside for testing whereas

Table 1. Confusion matrix of the blockwise classification of lung tissue patterns using a leave–one–patient–out cross–validation in %. Global arithmetic and geometric means of 75.1% and 74.7% are obtained respectively. N_{vox} denotes the number of manually segmented voxels used for evaluation and N_{cases} the number of patients.

	healthy	emphysema	ground glass	fibrosis	micronodules	N_{vox}	N_{cases}
healthy	**78.1**	2.8	0.7	0.2	18.1	63'914	7
emphysema	0.9	**70.1**	0	4.7	24.2	61'578	5
ground glass	4.6	1.6	**76**	14.7	3.1	644'814	21
fibrosis	2.3	1.9	17	**73.5**	5.3	860'474	28
micronodules	13.7	1.8	2.2	6.7	**75.7**	1'436'055	10

Table 2. Performance measures of the blockwise classification of the lung tissue patterns using QWF and GLH features

	recall	precision	F–measure	accuracy
healthy	78.4	78.2	78.3	91.3
emphysema	89.6	70.2	78.7	92.4
ground glass	79.2	76	77.6	91.2
fibrosis	73.6	73.5	73.6	89.4
micronodules	59.9	75.6	66.8	85

Table 3. Mean precisions based on the diagnosis of the retrieved cases. The values of the weight a_i that allowed best global precisions show the respective importances of the modalities. Abbreviations: BOOP: bronchiolitis obliterans organizing pneumonia, PCP: pneumocystis pneumonia.

	$P@1$	$P@5$	$P@10$	$P@N_r$	N_r
Fibrosis	79.2	58.3	51.7	42.7	24
BOOP	60	20	18	20	5
Miliary tuberculosis	71.4	48.6	34.3	42.9	7
PCP	25	20	10	25	4
Hypersensitivity pneumonitis	54.5	40	39.1	38	11
Acute interstitial pneumonia	66.7	33.3	25.5	27.2	9
Sarcoidosis	100	66.6	52.2	56.8	9
average/total	59.4	39.7	34.2	32.4	69
weights $a_{1,2,3}$ in (4)	8;1;39	6;9;38	8;5;48	10;4;41	

the remaining images are used to train the SVMs. For each case, lung volumes are segmented using YaDiV, where a tab was added for the closing operation. An example of the segmentation is depicted in Figure 1.

Then, the whole lung volume is segmented using a distance between the centers of the blocks equal to 4 pixels, leading to an overlap of 87.5% (see Figure 2). Table 1 shows the confusion matrix of the segmented tissue sorts. The associated performance measures are listed in Table 2. Note that some patients may contain several sorts of lung tissue. To assess case–based retrieval performance, mean precisions at rank 1, 5, 10 and at rank equal to the number of instances of the diagnosis N_r are computed using a leave–one–patient–out cross–validation with 69 cases (see Table 3). A grid search for optimal weights of the modalities in (4) is carried out with $a_j \in [0:50[$.

4   Discussion

The results obtained with the various components of the proposed CAD are discussed in this section. Our experience with the segmentation of 69 lung volumes shows that the 3D region growing associated with closing allows an almost fully–automatic segmentation. However, the trachea is included as lung tissue in most cases, which may bias the volume percentages of the five patterns in (4). Manual corrections are required when the closing operation cannot fill large regions of consolidated tissue.

The automatic segmentation of the lung tissue is a crucial step for the success of the CAD. The accuracies obtained in Table 2 show that the SVM classifier can learn efficiently from the hybrid feature space. However, the recurrent confusion between *healthy* and *micronodules* patterns suggests that the decision boundaries are not trivial in some cases (see also Figure 2). Table 1 also shows recurrent confusions between *fibrosis* and *ground glass*. This may be partially explained by the fact that *fibrosis* patterns are most often accompanied with small regions of *ground glass* because of the re–distribution of the perfusion to the functional tissue remaining. This has the effect to overload the healthy tissue which thus has the visual appearance of *ground glass* because of increased attenuation. However, during the annotations sessions, the label *fibrosis* was assigned to the whole ROI leading to classification errors when the system correctly detects the small *ground glass* regions. Using the clinical context of the images such as the age of the patient showed to allow clarifications between visually similar patterns in [18]. For instance, *micronodules* in a 20–year–old subject are very visually similar to *healthy* tissue surrounded by vessels of a 80–year–old man. In case of unbalanced classes, SVMs classifiers with asymmetric margins can be used to favor minority classes. At the border of the lungs, misclassifications occur due to the response of the wavelets to the sharp change of intensity. A solution to this is to use the symmetry of the lung tissue using the tangent to the lung border as axis. To remove noise in the blockwise classification, a 3D averaging of the lung regions may avoid small isolated regions.

The retrieval precisions presented in Table 3 are currently fairly low to be used in clinical routine but show the feasibility of indexing ILD cases using the volumes of automatically segmented lung tissue as well as clinical parameters. It is important to note that the link between visual similarity of two HRCT scans and their associated diagnoses is not straightforward. The values of the weights $a_{1,2,3}$ that allowed best performances reflect the importance of the each modality. High values obtained for a_3 shows the unexpectedly high importance of the clinical parameters of secondary priority. High variations of the precision can be explained by the fact that the number of cases is still fairly small, particularly for BOOP and PCP. The link between visual similarity of two HRCT scans and their associated diagnosis is not straightforward. Further improvements are required to highlight the importance of the visual similarity: the low–level feature vector v can be used directly as in Equation (4), under the condition to overcome the difficulty in setting up a standardized localization system for the lung anatomy.

5 Conclusion and Future Work

Image-based diagnosis aid tools for ILDs are available for evaluation to clinicians at the Emergency Radiology Service of the HUG. A web–based graphical interface is available to submit visual and textual queries. The recognition rate is obtained with an experimental setup that is similar to actual clinical situations. By analyzing every slice of the image series, it minimizes the risk of missing important lesions. Future work is required to reduce false detections of *micronodules* as well as to improve the precision of case–based retrieval.

Acknowledgments

This work was supported by the Swiss National Science Foundation (FNS) with grant 200020–118638/1, the equalization fund of Geneva University Hospitals and University of Geneva (grant 05–9–II) and the EU 6th Framework Program in the context of the KnowARC project (IST 032691).

References

1. Aziz, Z.A., Wells, A.U., Hansell, D.M., Bain, G.A., Copley, S.J., Desai, S.R., Ellis, S.M., Gleeson, F.V., Grubnic, S., Nicholson, A.G., Padley, S.P., Pointon, K.S., Reynolds, J.H., Robertson, R.J., Rubens, M.B.: HRCT diagnosis of diffuse parenchymal lung disease: inter–observer variation. Thorax 59(6), 506–511 (2004)
2. Sluimer, I.C., Schilham, A., Prokop, M., van Ginneken, B.: Computer analysis of computed tomography scans of the lung: a survey. IEEE Transactions on Medical Imaging 25(4), 385–405 (2006)
3. Shyu, C.R., Brodley, C.E., Kak, A.C., Kosaka, A., Aisen, A.M., Broderick, L.S.: ASSERT: A physician–in–the–loop content–based retrieval system for HRCT image databases. Computer Vision and Image Understanding (special issue on content–based access for image and video libraries) 75(1/2), 111–132 (1999)
4. Uppaluri, R., Hoffman, E.A., Sonka, M., Hunninghake, G.W., McLennan, G.: Interstitial lung disease: A quantitative study using the adaptive multiple feature method. American Journal of Respiratory and Critical Care Medicine 159(2), 519–525 (1999)
5. Fetita, C.I., Chang-Chien, K.-C., Brillet, P.-Y., Prêteux, F., Grenier, P.: Diffuse parenchymal lung diseases: 3D automated detection in MDCT. In: Ayache, N., Ourselin, S., Maeder, A. (eds.) MICCAI 2007, Part I. LNCS, vol. 4791, pp. 825–833. Springer, Heidelberg (2007)

6. Webb, W.R., Müller, N.L., Naidich, D.P. (eds.): High–Resolution CT of the Lung. Lippincott Williams & Wilkins, Philadelphia (2001)
7. Kim, N., Seo, J.B., Lee, Y., Lee, J.G., Kim, S.S., Kang, S.H.: Development of an automatic classification system for differentiation of obstructive lung disease using HRCT. Journal of Digital Imaging 22(2), 136–148 (2009)
8. Dundar, M., Fung, G., Bogoni, L., Macari, M., Megibow, A., Rao, B.: A methodology for training and validating a cad system and potential pitfalls. In: CARS 2004 – Computer Assisted Radiology and Surgery. Proceedings of the 18th International Congress and Exhibition. International Congress Series, vol. 1268, pp. 1010–1014 (2004)
9. Müller, H., Clough, P., Hersh, B., Geissbühler, A.: Variation of relevance assessments for medical image retrieval. In: Marchand-Maillet, S., Bruno, E., Nürnberger, A., Detyniecki, M. (eds.) AMR 2006. LNCS, vol. 4398, pp. 232–246. Springer, Heidelberg (2007)
10. Müller, H., Michoux, N., Bandon, D., Geissbuhler, A.: A review of content–based image retrieval systems in medicine – clinical benefits and future directions. International Journal of Medical Informatics 73(1), 1–23 (2004)
11. Müller, H., Rosset, A., Garcia, A., Vallée, J.P., Geissbuhler, A.: Benefits from content–based visual data access in radiology. RadioGraphics 25(3), 849–858 (2005)
12. Caritá, E.C., Seraphim, E., Honda, M.O., Mazzoncini de Azevedo-Marques, P.: Implementation and evaluation of a medical image management system with content–based retrieval support. Radiologia Brasileira 41(5), 331–336 (2008)
13. Aisen, A.M., Broderick, L.S., Winer-Muram, H., Brodley, C.E., Kak, A.C., Pavlopoulou, C., Dy, J., Shyu, C.R., Marchiori, A.: Automated storage and retrieval of thin–section CT images to assist diagnosis: System description and preliminary assessment. Radiology 228(1), 265–270 (2003)
14. Unser, M.: Texture classification and segmentation using wavelet frames. IEEE Transactions on Image Processing 4(11), 1549–1560 (1995)
15. Van De Ville, D., Blu, T., Unser, M.: Isotropic polyharmonic B–Splines: Scaling functions and wavelets. IEEE Transactions on Image Processing 14(11), 1798–1813 (2005)
16. Depeursinge, A., Van De Ville, D., Unser, M., Müller, H.: Lung tissue analysis using isotropic polyharmonic B–spline wavelets. In: MICCAI 2008 Workshop on Pulmonary Image Analysis, New York, USA, Lulu, September 2008, pp. 125–134 (2008)
17. Smeulders, A.W.M., Worring, M., Santini, S., Gupta, A., Jain, R.: Content–based image retrieval at the end of the early years. IEEE Transactions on Pattern Analysis and Machine Intelligence 22(12), 1349–1380 (2000)
18. Depeursinge, A., Racoceanu, D., Iavindrasana, J., Cohen, G., Platon, A., Poletti, P.A., Müller, H.: Fusing visual and clinical information for lung tissue classification in HRCT data. Journal of Artificial Intelligence in Medicine (to appear, 2009)

Image Retrieval for Alzheimer's Disease Detection

Mayank Agarwal and Javed Mostafa

Biomedical Research and Imaging Center
Laboratory of Applied Informatics Research
University of North Carolina at Chapel Hill, Chapel Hill, U.S.A
magarwal@email.unc.edu, jm@email.unc.edu

Abstract. A project is described with the aim to develop a Computer-Aided Retrieval and Diagnosis of Alzheimer's disease. The domain of focus is Alzheimer's disease A manually curated MRI data set from the Alzheimer's Disease Neuroimaging Initiative (ADNI) project (http://www.loni.ucla.edu/ADNI/) was used for training and validation. The system's main function is to generate accurate matches for any given visual or textual query. The system gives an option to perform the matching based on a variety of feature-sets, extracted using an adaptation of a discrete cosine transform algorithm. Classification is conducted using Support Vector Machines. Finally, ranking of most accurate matches are generated by applying an Euclidean distance score. The overall system architecture follows a multi-level model, permitting performance analysis of components independently. Experimental results demonstrate that the system can produce effective results.

1 Motivation

Researchers have been actively involved in understanding the structural changes that occur in volume and architecture of the brain. Research has shown that these changes are often associated with various neurological disorders such as Alzheimer's disease and schizophrenia. Many advances have been made in accurately describing these changes with promising results[1,2]. The inability to search over large image database due to limited or no associated textual information emphasizes the need to base retrieval on visual characteristics. However, not many systems exist that use image features for the purpose of retrieval especially when we consider the field of medicine. Studies have shown that the performance of clinicians and radiologists have improved significantly with the application of image retrieval systems particularly for clinicians new to the field[3].

Current medical imaging systems can produce a vast array of imaging data in a relatively short span of time. But only a few systems provide functions for automatically organizing and indexing images, and systems that offer effective retrieval and analysis functions are rare. Our goal is to conduct classification and indexing of medical images in near real-time, i.e. soon after the capture process is completed. A related goal is to develop high-accuracy retrieval and analysis functions over a dynamic image collection which is continuously expanding.

B. Caputo et al. (Eds.): MCBR-CDS 2009, LNCS 5853, pp. 49–60, 2010.

2 Problem Description

There is a difficult trade-off involved in designing retrieval systems that generate accurate results and produce the results in an efficient manner. The challenge lies in achieving a balance between the image representation process and access-latency. Generally, a representation process improves as more time is dedicated to it; however, it leads to increased access-latency.

We created a research platform for medical image retrieval and to analyze new and different approaches for reaching a balance in the trade-off. The platform, called ViewFinder Medicine (viewfM), was designed to support image manipulation by specialists in the domain of Alzheimer's disease. We designed viewfM based on a multi-level architecture, permitting convenient integration of modules for feature generation, classification, and ranking. As a training collection we used the pre-classified image data set from the Alzheimer's disease Neuroimaging Initiative (ADNI) at UCLA. The main theme of this paper is to identify the tradeoff points across certain dimensions:

1. representation vs classification
2. representation vs retrieval performance
3. classification vs retrieval performance

In the following sections we describe our system architecture, implementation of critical algorithms and experimental results.

3 Architecture

Our approach to the problem extends the multi-level retrieval model proposed by Mostafa, et al.[4]. Multi-level approach increases flexibilty for implementing key components as different algorithms can be supported. Components critical to the system are shown in Fig.1. as a multi-level architecture. The details of the steps involved in content representation and access are discussed next. In the following sections we describe the three primary steps in the retrieval process: segmentation and feature extraction, classification, and ranking.

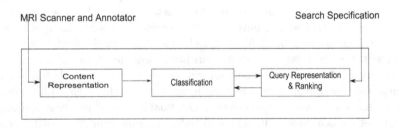

Fig. 1. Multi-level IR Approach to Image Retrieval

3.1   Segmentation and Feature Extraction

MRI modality may be impacted by several factors that may influence the search process. MRI process is subject to variations and may impact image characteristics such as intensity, orientation, shape, etc. Artifacts may be taken care of by labeling the image and later thresholding it. External noise in the scans is eliminated by first finding an appropriate circle to fit the brain structure and later extracting the structure representing the brain along with the skull. To help with the subsequent steps only the needed information in the brain was utilized for further analysis by finding a bounding box around the image.

An MRI scan is segmented into gray matter, white matter and cerebrospinal fluid. From the perspective of a radiologist, information pertinent to Alzheimer's disease is present in the hippocampus region in both the gray matter and the white matter. Kelemen in his work on automatic 3-D segmentation used shape and boundary information [5]. To segment the images we created a histogram of the number of voxels belonging to each gray level in the image to obtain the optimum levels at which to threshold the image. The optimum level corresponds to the boundary between the gray matter, white matter and the fluid.

Features are an important part of the content based retrieval process. "What features will aid the retrieval better" has been the center of contention for a long time. Variations in texture can be used to convey information about the existence of blotches in the brain which are typical of Alzheimer's disease. Research has shown texture analysis conducted on spatial domain represented as Gray Level Co-occurence matrix, features produced using Discrete Cosine Transformation (DCT) have superior discrimination power [6,7,8]. Additionally, DCT has high energy compaction and preserves image data correlation [9]. DCT has been extensively used in JPEG compression.

High dimensional data can make the computational complexity a limiting factor. Based on discussions with physicians, we identified a subset of slices which are highly likely to indicate Alzheimer's disease. On a slice by slice basis 2-D DCT was applied on the segmented gray matter, white matter and cerebrospinal fluid independently to get a 2-dimensional matrix of coefficients arranged in decreasing order of their spatial frequency. Normalized average of the three generated matrix was thus taken to produce a single 2-D matrix for the slice. Subsequently each slice information was transformed into a row of values and all such rows for were combined to create a 2-D matrix representing the texture of the scan. The features generated were stored in the backend database. To facilitate faster retrieval, a generalized inverted index was created on 2-D features.

3.2   Classification

When a user issues a query which is a MR scan, the system automatically classifies the query image into one of the three classes namely, AD, Mild cognitive impairment (MCI), and a normal control. The classification process is based on the DCT features described earlier and uses Support Vector Machines (SVM) as the classifier. SVM is a hyperplane classifier and reduces the problem to finding an optimal hyperplane that maximizes the difference between the classes. SVM classification has been applied in a wide array of fields [10,11]. Recently, SVM classification has been applied in medical imaging as well [12,13,14]. We used a generalized multi class SVM classifier for classification.

3.3 Ranking

Up on classification into a class, the distance between the input image and all the images belonging to that class was calculated. The distance which is Euclidean in nature is based on the 2-D features stored in the database.

$$\sum_{i=1}^{m}\sum_{j=1}^{n}(x_{ij} - y_{ij})^2$$

The distance for an m×n matrix is computed using the above equation. The images were then ranked according to the computed distance. Classification provides us with decision values which represent the probability that an image belongs to a particular class. The probability value was combined with the distance value according to the formula below, thus

$$R = p \times \sum_{i=1}^{m}\sum_{j=1}^{n}(x_{ij} - y_{ij})^2$$

generating the ranking for each image, where p represents the decision value or the probability generated by the classifier that the input image belongs to a particular class. In Fig.2.-3. we show the viewfM user interface with retrieved results displayed in ranked order.

4 User Interaction

User interface is one of the key component of viewfM. ViewFinder supports textual keyword based queries and visual promote searches on images. Fig. 2.. shows how a keyword based search progresses through viewfinder. In Fig.2.(a) the user is presented with a random set of images covering all the classes arranged in a fish-eye view. The system is self guiding as it builds on user efforts to input the query and provides auto completion choices. In Fig.2.(b) the ranked results are presented to the users.

Promote search can also be used to initiate the information seeking task. Fig.3. show another possible path to search for relevant scans. The user is presented with the deafult layout where the user after looking at the subject information, selects one of the closest scan and conducts a visual promote. As a result the chosen image is moved to the center of viewfM (see Fig.3.(b)), and other similar images based on visual characteristics and the classification results are arranged around this image in a ranked fashion. The user can promote or conduct a keyword based search to further refine the results.

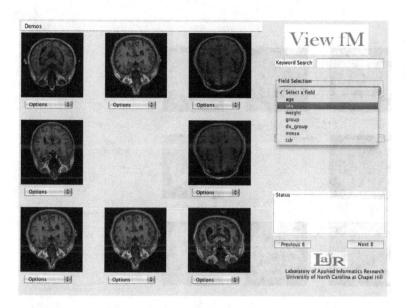

(a) User is presented with a drop down menu to choose available metadata fields to search on

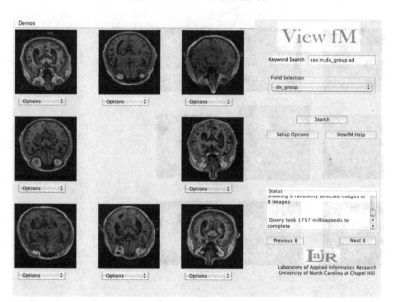

(b) Search results are presented to the user in fish-eye view

Fig. 2. Flow of a textual query

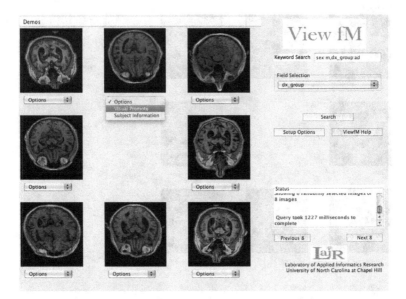

(a) User chooses promote search on one of the results in Fig.2.(b)

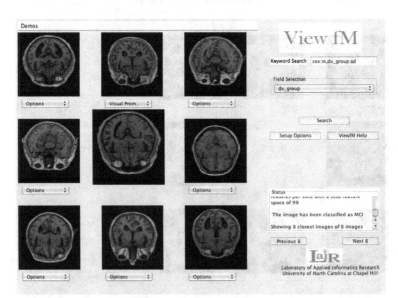

(b) Search results are presented to the user

Fig. 3. Flow of a visual search

5   Experimental Setup

For the purpose of evaluation, a subset of manually pre-classified MRI scans from Alzheimer's Disease Neuroimaging Initiative was used. The dataset contains MR scans of 354 cases having a total of 1020 scans with an average of 2.88 scans per subject. All the scans are T-1 weighted contrast enhanced MP-RAGE images in ANALYZE format.

5.1   Query Distribution

Among the 1020 scans a subset of 25 images was chosen randomly as a representative sample. A uniform distribution was followed between the classes to represent the query topics. The query distribution across the class for AD, MCI, and Normal are 9, 8, and 8 respectively as summarized in Table 1.

Table 1. Distribution of Queries across space

Query Topic	Number of Queries
AD	9
MCI	8
Normal	8

5.2   Methodology

The experimental methodology was designed to establish the trade-off points between optimum efficiency and effectiveness factors. In this context we first compare the accuracy of the retrieval process without a classifier in the system with the accuracy of the system with a classifier. The accuracy of retrieval was measured in terms of precision and recall. Precision indicates the proportion of relevant images retrieved when compared to the entire set of the collection, while recall represents the proportion of the relevant images retrieved when compared to the actual set of relevant images that should have been found. The retrieval performance was subsequently tested with an SVM classifier when a notable improvement in retrieval performance was observed.

We also conducted experiments to examine the parameters chosen for representation. These include a centrality measure and window sizes. Centrality measure is an indicator of how many slices we choose to represent the scans. Window size determines how many discrete cosine transform co-efficients are utilized as features.

A final set of experiments was conducted with a focus on capturing the effect of ranking on classification error. These were based on the hypothesis that the combined effect of multiple levels, can minimize the impact of errors contributed by the individual levels. A set of related experiments focus on combining multiple classes while ranking to highlight any inter-class variability that may influence the retrieval performance.

6   Results

In our first experiment we measured the retrieval performance of the system with minimal representation and no intermediate classification. To calculate performance we chose to use average precision at different cutoffs. Precision at cutoff indicates the precision after a certain number of images are retrieved. The user at any time is presented with a "matrix" of eight images, with top-left image being the most relevant, and other images in descending relevance order. The average precision was calculated over 25 queries.

Table 2. Average precision without classification with a centrality parameter of 11 and windowsize of 5×5

Avg Precision	Without Classification
at 10	0.420
at 20	0.380
at 30	0.370
at 40	0.334
at 50	0.332

Table 2. suggests that the accuracy of the retrieval is subpar without a classifier.

Performing the experiments on the queries by varying the window size and the centrality parameter, maximum accuracy of 72.6% was obtained for classification. From the Table 3. below it is evident that after a certain threshold, adding more information to the system does not help with classification and indeed decreases the performance of the system.

Table 3. Classification Accuracy

	3×3	5×5	7×7	9×9
11 slices	69.7279	71.7687	70.068	70.7483
21 slices	70.068	71.0884	71.4286	70.068
151 slices	68.7075	71.0884	70.068	69.7279

To look deeper into the classification performance we examined the classification accuracy for each class (see Fig. 4.). The maximum classification accuracy was obtained for Mild Cognitive Impairment of all the classes at 73.6%. The result slightly contradicts the recent published results [15]. It, however, supports the assumption that more localized the atrophy is, the easier it is to classify the scan into one of the three categories.

The goal of the next experiment was to determine if the classification level can assist in improving the retrieval performance. A more specific scenario we examined was for a randomly selected set of queries find the best ranking for

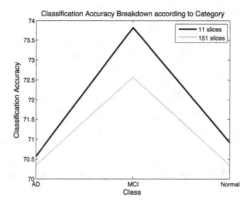

Fig. 4. Classification Accuracy per Class

Table 4. Average precision with classification with a centrality parameter of 11 and windowsize of 5×5

Avg Precision	With Classification
at 10	0.816
at 20	0.783
at 30	0.776
at 40	0.750
at 50	0.730

each query (i.e. find the closest matching patients). Using our multilevel approach with a classifier, the results we achieved are summarized in the Table 4.

The retrieval performance of the system with classification in contrast to without classification appeared to be significantly better. This can be attributed to the fact that the intermediate classifier reduces the search space to only a single class which will either be relevant or non-relevant.

From Fig.5.(a), it can be concluded that as the cutoff level drops the ranking becomes more precise. Hence, most of the relevant results are found at the start of the interaction cycle - a desirable outcome we wished to achieve, given the manner in which retrieved results are displayed (see Fig.3.).

In looking further into the classification process, we determined the decision values can further be used to enhance the retrieval beyond just generating the closest class. The decision values provided by the classifier can be combined with the distance to come up with a similarity measure. When considering only a single class case, this does not affect the ranking of the images. However, combining multiple classes for ranking in the above fashion improves the ranking and the retrieval performance. As the results of Fig.5.(b)-(d) demonstrate, the effect of error in classification can be somewhat reduced by applying a practical ranking measure which combines distance with classification prediction.

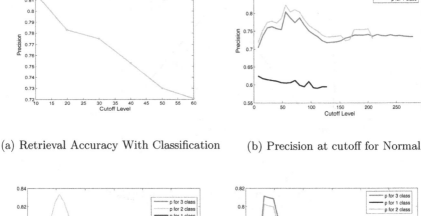

(a) Retrieval Accuracy With Classification   (b) Precision at cutoff for Normal

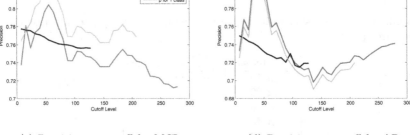

(c) Precision at cutoff for MCI   (d) Precision at cutoff for AD

Fig. 5. Classwise Average Precision at Cutoff with 1 class, 2 class and 3 class combined

7   Conclusion

The field of medical image retrieval is still in its early stages. This study developed a platform to find similar MR images which can be used by radiologists to assist them in diagnosing and finding information about other people with similar conditions. We demonstrated that in a multi-level architecture, combining classification and ranking, superior results can be achieved, whereby computational resources dedicated to the presentation can be balanced with the demand for supporting responsive user interactions.

A more in-depth analysis which focuses on time as an efficiency factor and measures representation and ranking performance based on different levels of feature selections is needed. A related issue that requires closer investigation is a usability evaluation based on realistic scenarios. This may involve identifying a target user group willing to take time to review the images to subjectively judge the performance of the system to create a benchmark. Continued studies on different features, interactions with radiologists and other clinicians, and incorporation of relevance feedback will be critical for future research in this field.

Acknowledgement

We thank the UNC Biomedical Research and Imaging Center (BRIC) and Translational and Clinical Sciences (TraCS) Institute for their support. Data collection and sharing for this project were funded by the Alzheimers Disease Neuroimaging Initiative (ADNI; Principal Investigator: Michael Weiner; NIH grant U01 AG024904).

References

1. Akselrod-Ballin, A., Galun, M., Gomori, M., Basri, R., Brandt, A.: Atlas guided identification of brain structures by combining 3D segmentation and SVM classification. In: Larsen, R., Nielsen, M., Sporring, J. (eds.) MICCAI 2006. LNCS, vol. 4191, pp. 209–216. Springer, Heidelberg (2006)
2. Jack Jr., C., Bentley, M., Twomey, C., Zinsmeister, A.: MR imaging-based volume measurements of the hippocampal-formation and anterior temporal-lobe-validation studies. Radiology 176, 205–209 (1990)
3. Aisen, A.M., Broderick, L.S., Winer-Muram, H., Brodley, C.E., Kak, A.C., Pavlopoulou, C., Dy, J., Shyu, C.R., Marchiori, A.: Automated Storage and Retrieval of Thin-section CT Images to Assist Diagnosis: System Description and Preliminary Assessment. Radiology 228(1), 265–270 (2003)
4. Mostafa, J., Mukhopadhyay, S., Palakal, M., Lam, W.: A multilevel approach to intelligent information filtering: model, system, and evaluation. ACM Trans. Inf. Syst. 15(4), 368–399 (1997)
5. Kelemen, A., Szekely, G., Gerig, G.: Elastic model-based segmentation of 3-D neuroradiological data sets. Med. Img. 18(10), 828–839 (1999)
6. Huang, Q., Dony, R.: Neural network texture segmentation in equine leg ultrasound images. In: Canadian Conference on Electrical and Computer Engineering, vol. 3, pp. 1269–1272 (2004)
7. Sorwar, G., Abraham, A., Dooley, L.: Texture Classification Based on DCT and Soft Computing. In: FUZZ-IEEE, pp. 545–548 (2001)
8. Ngo, C.W., Pong, T., Chin, R.T.: Exploiting Image Indexing Techniques in DCT Domain. In: IAPR International Workshop on Multimedia Media Information Analysis and Retrieval, pp. 1841–1851 (1998)
9. Singh, P.K.: Unsupervised segmentation of medical images using dct coefficients. In: VIP 2005: Proceedings of the Pan-Sydney area workshop on Visual information processing, pp. 75–81. Australian Computer Society, Inc., Australia (2004)
10. Li, J., Allinson, N., Tao, D., Li, X.: Multitraining Support Vector Machine for Image Retrieval. IEEE Transactions on Image Processing 15(11), 3597–3601 (2006)
11. Pontil, M., Verri, A.: Support vector machines for 3D object recognition. IEEE Transactions on Pattern Analysis and Machine Intelligence 20, 637–646 (1998)
12. Wang, C.M., Mai, X.X., Lin, G.C., Kuo, C.T.: Classification for Breast MRI Using Support Vector Machine. In: CITWORKSHOPS 2008: Proceedings of the 2008 IEEE 8th International Conference on Computer and Information Technology Workshops, Washington, DC, USA, pp. 362–367. IEEE Computer Society, Los Alamitos (2008)

13. Du, X., Li, Y., Yao, D.: A Support Vector Machine Based Algorithm for Magnetic Resonance Image Segmentation. In: ICNC 2008: Proceedings of the 2008 Fourth International Conference on Natural Computation, Washington, DC, USA, pp. 49–53. IEEE Computer Society, Los Alamitos (2008)

14. Mourao-Miranda, J., Bokde, A.L., Born, C., Hampel, H., Stetter, M.: Classifying brain states and determining the discriminating activation patterns: Support Vector Machine on functional MRI data. NeuroImage 28(4), 980–995 (2005); Special Section: Social Cognitive Neuroscience

15. Kloppel, S., Stonnington, C.M., Chu, C., Draganski, B., Scahill, R.I., Rohrer, J.D., Fox, N.C., Jack, C.R., Ashburner, J., Frackowiak, R.S.J.: Automatic classification of MR scans in Alzheimer's disease. Brain 131(3), 681–689 (2008)

Statistical Analysis of Gait Data to Assist Clinical Decision Making

Nigar Şen Köktaş[1] and Robert P.W. Duin[2]

[1] Department of Mathematics and Computer Sciences,
Faculty of Arts and Sciences, Çankaya University,
06530 Balgat, Ankara, Turkey
nigarsen@cankaya.edu.tr
[2] Faculty of Electrical Engineering, Mathematics and Computer Sciences,
Delft University of Technology, The Netherlands
r.duin@ieee.org

Abstract. Gait analysis is used for non-automated and automated diagnosis of various neuromuskuloskeletal abnormalities. Automated systems are important in assisting physicians for diagnosis of various diseases. This study presents preliminary steps of designing a clinical decision support system for semi-automated diagnosis of knee illnesses by using temporal gait data. This study compares the gait of 111 patients with 110 age-matched normal subjects. Different feature reduction techniques, (FFT, averaging and PCA) are compared by the Mahalanobis Distance criterion and by performances of well known classifiers. The feature selection criteria used reveals that the gait measurements for different parts of the body such as knee or hip to be more effective for detection of the illnesses. Then, a set of classifiers is tested by a ten-fold cross validation approach on all datasets. It is observed that average based datasets performed better than FFT applied ones for almost all classifiers while PCA applied dataset performed better for linear classifiers. In general, nonlinear classifiers performed quite well (best error rate is about 0.035) and better than the linear ones.

Keywords: gait analysis; statistical pattern classifiers; clinical decision support systems.

1 Introduction

Gait Analysis is defined as the analysis of walking patterns of humans. A major application area is in the clinical decision-making and treatment processes for neuro-muskuloskeletal diseases, among others such as security clearance systems and human identification [1]. Increasing number of Gait Laboratories in hospitals and respectively growing amounts of collected quantitative data necessitates the use of automated systems.

The use of pattern recognition techniques for analyzing gait data is studied before with varying success. Neural Networks [2-7] and Support Vector machines [7-9] are among the approaches used. Most of these works concentrates on a binary decision of the presence or non-presence of a disease. The number of subjects used for training is

B. Caputo et al. (Eds.): MCBR-CDS 2009, LNCS 5853, pp. 61–68, 2010.

generally quite low (About 10 or 15), with questionable reliability of the results. Used features are measured quantitative data such as joint angles, force plate data, cadence (distance between toes in one step) etc.

There are some significant difficulties in the design of automated systems because of the multidimensional and complex structure of the gait data. Feature reduction and selection always become an important part of the gait analysis studies. The most commonly used method for the feature extraction is based on the estimation of parameters (peak values, ranges) as descriptors of the gait patterns. In that case the classification is done according to the differences between the class averages of the training set and the parameters of the new subjects [10,11]. This method is subjective and neglects the temporal information of the gait data. There are examples of feature reduction techniques for gait data, such as Fast Fourier Transform (FFT) [3, 4, 11], Principal Component Analysis (PCA) [10-12], and wavelet transform [2]. Since all these studies differ from each other in description of the gait variables (such as subject type, measurement tools, type of variables, anatomical levels), and in construction of the classifiers, the results do not give a proper indication of the comparative performances.

Automated selection of the gait parameters is not observed in gait classification literature; medical experts select them or previous studies are taken as references. Actually, there are many medical practices, testing the variations in the gait parameters, which are caused by the related illness [12-14]. In [6] a gait classification study is performed to assist physician's decision making. Combination of MLPs is used for classification purpose, but no feature reduction or selection processes are applied. Therefore validity of the system depends on the expertise of the physician who selects features manually. Also, the judgments may vary in different experts leading the different interpretations of the classifiers. Obviously, automated selection lessens the dependence and the load on the experts and gives more freedom to the researchers.

The objective of the presented study is to propose a complete gait classification approach for automated diagnosis of the knee Osteoarthritis (OA) by comparing the convenient methods for preprocessing (feature reduction/selection) and classification. Most important contribution of the study is presenting a guide for further studies aiming classification of temporal and high dimensional medical data. It serves a base for design of clinical decision support systems by suggested automated feature reduction/selection methods and classification algorithms.

The following section discusses the details of gait analysis and data collection process. Section two gives more details about the used method and experiments, then discussion of the results are presented. Finally conclusion is presented.

1.1 Data Collection

Gait analysis enables the clinicians to differentiate gait deviations objectively. It serves not only as a measure of treatment outcome, but also as a useful tool in planning ongoing care of various neuromuskuloskeletal disorders such as cerebral palsy, stroke, OA, as an alternative or supportive to other approaches such as X-rays, MRI, chemical tests etc [13, 14]. Gait process is realized in a 'Gait Laboratory' by the use of computer-interfaced video cameras to measure the patient's walking motion. Electrodes called markers placed on the skin to follow the muscle activity by using the infrared domain of the cameras. In addition, force platforms embedded in a walkway monitor the forces and torques produced between the patient and the ground.

In this study, data are collected in Ankara University Faculty of Medicine Department of Physical Medicine and Rehabilitation Gait Laboratory by the gait analysis experts, from 110 normal and 111 OA patients. In this laboratory, there are standard gait laboratory equipments, which are supported by "VICON" a commercial system for gait analysis. A subject is asked to walk on the platform and in one cycle of gait, temporal changes of joint angles, joint moments, joint powers, force ratios and time-distance parameters from four anatomical levels (pelvis, hip, knee and ankle) and three motion planes (sagittal, coronal and traversal) of the subjects are gathered. Resultant data (such as knee angle/time) is tabulated in graphic/numerical forms. Figure 1 shows examples to temporal gait parameters.

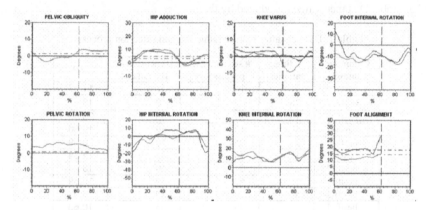

Fig. 1. Temporal gait parameters (data set B): The data is reported in 2-D charts with the abscissa defined as the percentage of the gait cycle and the ordinate displaying the gait parameter

Each of these parameters is represented by a graph that contains N samples taken in equally spaced intervals for one gait cycle. So the parameters for a given subject can be arranged as a M-dimensional vector X as below:

$X = [X^{(1)}, X^{(2)}, \ldots X^{(M)}]$ where
$X^{(i)} = [X^{(i)}_1, X^{(i)}_2, \ldots X^{(i)}_N]$

2 Experiments

The software in gait laboratory produces a dataset for each subject, including 33 gait parameters; each has 51 sample points in time. Therefore a [*33 (parameters) x 51 (time samples)*] dimensional array is created for each subject. Since the total number of the features is too large relative to the number of subjects, most of the commonly used classifiers will suffer from the curse of dimensionality [15, 16]. Different reduction techniques are applied for comparison. Averaging consecutive time samples creates six datasets having different dimensions. Moreover, FFT is applied to each waveform and one, five, ten or 25 FFT coefficients rather than 51 time samples represent each parameter. A PCA applied dataset is also added to this list as third reduction method. PCA is a procedure that transforms a number of possibly correlated variables into a smaller number of uncorrelated variables called principal components (*PCs*). PCA is applied to feature matrix that is created by combing all gait parameters and a subset of principal components is selected for classification. Since first 65 PCs contain 95% of the total variance explained by all PCs, they are included in final dataset namely *dPCA*.

Following the reduction step, Mahalanobis distance (*MD*) is used to extract informative and discriminative features for effective classification. MD is a distance measure based on correlations between variables by which different patterns can be identified and analyzed. It is mostly used for determining similarity of an unknown sample set to a known one. In this study, the feature set creating the larger distance for two classes is selected as most discriminative feature set. Individual performances of the each gait parameters to discriminate two classes are compared. Table 1 shows the properties of these datasets created by combining the selected *best* features and the Mahalanobis Distance values produced by them.

Table 1. Datasets after feature reduction and selection processes

Composed dataset	# time samples	# selected gait parameters	Dimension of the dataset	*MD*
best51d	51	1	51	14.57
best25d	25	2	50	18.27
best10d	10	5	50	23.55
best5d	5	10	50	19.80
best2d	2	25	50	20.39
best1d	1	33	33	14.58
bestfft25	25	2	50	13.24
bestfft10	10	5	50	15.59
bestfft5	5	10	50	18.03
bestfft1	1	33	33	10.63
dPCA	-	-	65	31.93

Table 2. List of used classifiers

Classifier	Description [17]
Logistic Linear Classifier (loglc)	Computation of the linear classifier for the dataset by maximizing the likelihood criterion using the logistic (sigmoid) function.
Support vector classifier (svc)	Optimizes a support vector classifier for the dataset by quadratic programming. The default classifier is linear.
Linear Bayes Normal Classifier (ldc)	Computation of the linear classifier between the classes of the dataset by assuming normal densities with equal covariance matrices.
Quadratic Bayes Normal Classifier (qdc)	Computation of the quadratic classifier between the classes of the dataset assuming normal densities.
Feed-forward neural net classifier (bpxnc)	A feed-forward neural network classifier with 1 hidden layer with 5 units in layer. Training is stopped after 100 epochs. Standard Matlab
Feed-forward neural net classifier (lmnc)	initialization is used. Training algorithm: "bpxnc" for back-propagation, "lmnc" for Levenberg-Marquardt
Automaric radial basis SVM (rbsvc)	A support vector classifier having a radial basis kernel
Parzen Classifier (parzenc)	Scaled datasets and log functions are used for better results. Computation of the optimum smoothing parameter for the Parzen classifier between the classes in the dataset.
Parzen density based classifier (parzendc)	For each of the classes in the dataset a Parzen density is estimated. For each class, feature normalization on variance is included in the procedure

In classification stage, a set of linear and nonlinear classifiers is tested by a ten-fold cross validation method. PRTools [17] is used for classifier construction. Total of nine classifiers are used by default parameters of the toolbox. Some adjustments are applied for better classification accuracy. Table 2 shows descriptions and adjustments of these classifiers, for more information see also [15-18].

Since density based classifiers (ldc, qdc, parzenc) suffer from a low numeric accuracy in the tails of the distributions, 'log' function is used to compute log- densities of them. This is needed for over trained or high dimensional classifiers. Almost zero-density estimates may otherwise arise for many test samples, resulting in a bad performance due to numerical problems. Classifiers are tested by cross-validation which is a methods for estimating generalization error based on *resampling*. Table 3 shows the classification error rates observed as a result of ten fold cross validation for all datasets presented in Table 1.

Table 3. Generalization error rates of classifiers

Datasets	Classifiers								
	loglc	svc	ldclog	qdclog	bpxnc	lmnc	rbscv	parzenlog	parzendc
best1d	0,083	0,080	0,066	0,086	0,071	0,073	0,047	0,098	0,093
best2d	0,097	0,088	0,062	0,110	0,069	0,070	**0,035**	0,070	0,062
best5d	0,088	0,083	0,050	0,121	0,063	0,068	0,044	0,055	0,076
best10d	0,084	0,086	0,046	0,130	0,056	0,070	0,085	0,086	0,077
best25d	0,115	0,079	0,068	0,158	0,065	0,070	0,116	0,089	0,103
best51d	0,207	0,138	0,091	0,206	0,111	0,155	0,143	0,142	0,152
best1ffd	0,110	0,098	0,082	0,104	0,095	0,110	0,074	0,096	0,096
best5fftd	0,152	0,124	0,083	0,113	0,100	0,117	0,102	0,076	0,095
best10fftd	0,127	0,119	0,082	0,132	0,100	0,125	0,095	0,112	0,164
best25fftd	0,148	0,098	0,110	0,132	0,115	0,143	0,103	0,129	0,106
dPCA	0,072	0,074	0,049	0,157	0,057	0,065	0,054	0,051	0,205

3 Discussion of Results

Comparing three reduction techniques using MD values, it is observed that averaged datasets perform better than the corresponding FFT based dataset. PCA applied dataset created largest distance value for classes, expectedly. Because, both PCA and MD methods are performed on covariance matrix of the input set. PCA is a well-known method to reduce feature dimensionality but not suitable for explicit feature selection like in current study. Because the contributions of the gait parameters to mapped feature set is not understandable. So it would be better to select features by some other methods and then to apply PCA. MD values are used to score the discriminatory ability of gait parameters, too. The number of appearance times of the parameters in datasets is listed below:

- KFlex (Knee Flexion): 7
- HMAbd (*Hip* Abduction Moment): 7

- KMFlex (Knee Flexion Moment): 5
- KRot (Knee Rotation): 4
- KMVal (Knee Valgus Moment): 4
- KPTot (Knee Power Total): 3

Four knee-related gait parameters are selected by an expert physician for classification in a previous study implemented by the same data [6]. These parameters have high scores in above list (KFlex, KMFlex, KMVal, KPTot). This supports the validation of automated feature selection process used in this study. Another mentionable result of feature selection process is that a hip related parameter has a score as high as knee related ones. The high discriminatory ability of this feature shows that knee OA causes high variation in hip abduction moments as much as in knee flexion moments of the patients.

Comparing the performances of the classifiers on the basis of the current number of subjects, it may be concluded that nonlinear classifiers performed quite well and better than the linear ones. We have also observed that high regularization prevents linear classifiers learning from more data. If we evaluate the success rates of the datasets by a standard baseline method, such as Linear Discriminant classifier (*ldclog* in Table 3), we see that PCA and averaging applied datasets performed better than the FFT applied ones. Actually among the linear classifiers *ldclog* performed as good as non-linear ones. Radial basis support vector classifier produced best generalization rate for almost all datasets.

4 Conclusion

The results of the experiments performed in this study are important for defining further studies about automatic diagnosis of gait disorders. Starting from the beginning of the study three different feature reduction techniques are compared first by the MD criterion and then by performances of the classifiers. It is observed that datasets created by FFT techniques produce worse results than the others. PCA applied dataset produces best result by almost all classifiers. Obviously, non-linear classifiers perform better than the linear ones with current number of subjects, but it can be suggested that considering the training costs of the algorithms, linear classifiers with a convenient regularization rate may be included in the further studies with *more* data.

It is also concluded that, temporal information of the waveforms is not so significant for the classifier performances. The severe difference between the performances of the datasets, best1d (all gait parameters with one time sample), and best51d (one gait parameter with all time samples) shows that including more gait parameter is more informative than including more time samples.

We detect a high match between currently selected features and the ones suggested by gait analysis expert. It can be concluded that our selection criterion approximates the expert knowledge and so contributes to the validation of the approach. The interpretation of the results of selection process reveals considerable information about effects of the illness. For example, expert physician concludes that motion of the hip angle is distorted as much as that of knee angle for sick subjects. Automatic feature selection may be preferred to find variation in all levels and motion planes of the subjects But still, the feature selection process may be improved by using more general criteria such as 1-Nearest Neighbor performance and also by using different search strategies than individual selection such as forward or backward selection.

These experiments showed us that statistical pattern recognition algorithms produce promising results for automated analysis of the gait data. The comparison of available statistical approaches for gait classification is expected to guide researchers working on this area for further studies.

Acknowledgements

This study is supported by The Scientific and Technological Research Council of Turkey (TUBITAK).

References

1. Simon, S.R.: Quantification of human motion: gait analysis—benefits and limitations to its application to clinical problems. Journal of Biomechanics 37, 1869–1880 (2004)
2. Chau, T.: A review of analytical techniques for gait data, Part 2: Neural Networks and Wavelet Methods. Gait Posture 13, 102–120 (2001)
3. Kohle, M., Merkl, D., Kastner, J.: Clinical gait analysis by neural networks: issues and experiences. In: Proceedings of the 10th IEEE Symposium on Computer-Based Medical Systems, p. 138 (1997)
4. Barton, J.G., Lees, A.: An application of neural networks for distinguishing gait patterns on the basis of hip-knee joint angle diagrams. Gait & Posture 5, 28–33 (1997)
5. Sen Koktas, N., Yalabik, N., Yavuzer, G.: An Intelligent Clinical Decision Support System for Analyzing Neuromuskuloskeletal Disorders. In: International Workshop on Pattern Recognition in Information Systems, pp. 29–37 (2008)
6. Sen Koktas, N., Yalabik, N., Yavuzer, G.: Ensemble Classifiers for Medical Diagnosis of Knee Osteoarthritis Using Gait Data. In: Proceeding of IEEE International Conference on Machine Learning and Applications (2006)
7. Begg, R., Kamruzzaman, J.: A Comparison of Neural Networks and Support Vector Machines for Recognizing Young-Old Gait Patterns. In: Proceeding of IEEE TENCON Conference (2003)
8. Begg, R., Palaniswami, M., Owen, B.: Support Vector Machines for Automated Gait Classification. IEEE Transactions on Biomedical Engineering 1, 52–65 (2005)
9. Salazar, A.J., De Castro, O.C., Bravo, R.J.: Novel approach for spastic hemiplegia classification through the use of support vector machines. In: Proceedings of the 26th Annual International Conference of the Engineering in Medicine and Biology Society (2004)
10. Dobson, F., Morris, M.E., Baker, R., Graham, H.K.: Gait classification in children with cerebral palsy: A systematic review. Gait and Posture 25, 140–152 (2007)
11. Chau, T.: A review of analytical techniques for gait data, Part 1: Fuzzy, statistical and fractal methods. Gait Posture 13, 49–66 (2001)
12. Deluzio, K.J., Astephen, J.L.: Biomechanical features of gait waveform data associated with knee Osteoarthritis: An application of principal component analysis. Gait and Posture 25, 86–93 (2007)
13. Gök, H., Ergin, S., Yavuzer, G.: Kinetic and kinematic characteristics of gait in patients with medial knee arthrosis. Acta Orthop Scand 2002 73(6), 647–652 (2002)

14. Kaufman, K., Hughes, C., Morrey, B., An, K.: Gait characteristics of patients with knee Osteoarthritis. Journal of Biomechanics 34, 907–915 (2001)
15. Jain, A.K., Duin, R.P.W., Mao, J.: Statistical Pattern Recognition: A Review. IEEE Transactions on Pattern Analysis and Machine Intelligence 22(1), 4–37 (2000)
16. Duda, R.O., Hart, P.E., Stork, D.G.: Pattern Classification. John Wiley and Sons, New York (2001)
17. Duin, R.P.W.: PRTOOLS (version 4). A Matlab toolbox for pattern recognition. Pattern Recognition Group, Delft University of Technology (February 2004)
18. Kuncheva, L.I.: Combining Pattern Classifiers: Methods and Algorithms. Wiley-Interscience, Hoboken (2004)

Using BI-RADS Descriptors and Ensemble Learning for Classifying Masses in Mammograms

Yu Zhang, Noriko Tomuro, Jacob Furst, and Daniela Stan Raicu

College of Computing and Digital Media
DePaul University, Chicago, IL 60604, USA
{jzhang2,tomuro,jfurst,draicu}@cs.depaul.edu

Abstract. This paper presents an ensemble learning approach for classifying masses in mammograms as malignant or benign by using Breast Image Report and Data System (BI-RADS) descriptors. We first identify the most important BI-RADS descriptors based on the information gain measure. Then we quantize the fine-grained categories of those descriptors into coarse-grained categories. Finally we apply an ensemble of multiple Machine Learning classification algorithms to produce the final classification. Experimental results showed that using the coarse-grained categories achieved equivalent accuracies compared with using the full fine-grained categories, and moreover the ensemble learning method slightly improved the overall classification. Our results indicate that automatic clinical decision systems can be simplified by focusing on coarse-grained BI-RADS categories without losing any accuracy for classifying masses in mammograms.

Keyword: Mass Classification, BI-RADS, CADx.

1 Introduction

Breast cancer is the second leading cause of cancer related deaths for women in the U.S. after lung cancer [1]. At present, mammography screening is the most effective method for the early detection of breast cancer. However, the error rate of mammography screening is still high [2]. Many Computer-Aided Diagnosis (CADx) systems have been developed as a second opinion to assist radiologists [3].

Breast Image Report and Data System (BI-RADS) [4] is a set of lexicons describing breast lesions, which was developed by the American College of Radiology (ACR) to standardize the terminology in mammogram reports. The BI-RADS has been used in various research studies for the diagnoses of breast cancer. Kim et al. [5] used BI-RADS-based features, and applied a Support Vector Machine based on recursive feature elimination (SVM-RFE) for classifying abnormalities in mammogram images. Elter et al. [6] presented two CAD systems which use decision-tree learning and case-based reasoning for the prediction of breast cancer from BI-RADS attributes.

The research presented in this paper is part of an ongoing project for developing an image-based CADx system to classify suspicious masses in mammograms as malignant or benign. By studying BI-RADS descriptors, we will be able to identify

B. Caputo et al. (Eds.): MCBR-CDS 2009, LNCS 5853, pp. 69–76, 2010.
© Springer-Verlag Berlin Heidelberg 2010

important domain knowledge and effective methods to guide our image-based CADx system. For radiologists, the shape and margin of masses are two important descriptors to distinguish malignant from benign masses [7]. Mass shape and margin feature are both defined by five categories in BI-RADS. In this research, we hope to simplify the decision process of classifying suspicious masses by using the coarse-grained BI-RADS categories, and still achieve equivalent or higher classification accuracies with the ensemble learning method. By applying the same methods from this study, the technical aspect involved in an image-based system could be simplified without sacrificing any classification accuracy.

Figure 1 below depicts the schematic framework of our approach. First the BI-RADS descriptors are extracted from the overlay files which contain keywords that describe each abnormality; next feature selection is applied to identify the important descriptors, and the fine-grained categories of those descriptors are collapsed and converted into coarse-grained categories; then the dataset is split into subsets based on the coarse-grained categories; finally, an ensemble of classifiers (Decision Tree, Bayes Network, Neural Network, Support Vector Machine, and K-Nearest Neighbor) is formed for each data subset, and the one which produced the highest accuracy is selected. The final classifications for the whole dataset are obtained by combining the classifications derived by the individual classifiers. The results showed that, by using the BI-RADS shape descriptor with coarse-grained categories along with the margin descriptor and patient age feature, our ensemble learning method achieved the overall accuracy of 84.43% for classifying masses in mammograms as malignant or benign.

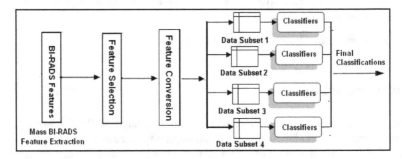

Fig. 1. Overall Framework of Our Approach

This paper is organized as follows: Section 2 reviews the BI-RADS descriptors and dataset, Section 3 describes the feature selection and data splitting methods, Section 4 presents the experimental results, and Section 5 discusses the conclusions and future work.

2 Data and BI-RADS Descriptors

2.1 Dataset Description

In this work, all mass instances were collected from the Digital Database for Screening Mammography (DDSM) from the University of South Florida [8], which is the largest publicly available resource for the mammogram analysis research community. In DDSM images, suspicious regions of interest (ROIs; including masses and micro-calcifications) are marked by experienced radiologists, and BI-RADS information is also annotated for each abnormal region. In our experiment, we used mass instance images digitized by LUMYSIS. We removed instances with mixed BI-RADS de-scriptors or images with extreme digitization artifacts – that left us with a total of 681 mass instances, where 314 were benign and 367 were malignant. The BI-RADS de-scriptors were extracted from the overlay files using Matlab, and all classifications were conducted using a Machine Learning tool called Weka [9], with 10-fold cross-validation.

2.2 BI-RADS Descriptors

In BI-RADS, mass shapes are defined as either round, oval, lobulated, irregular or architectural distortion. Usually a poorly defined shape is more likely to be malignant than a well-circumscribed mass [7]. Margin is the border of a mass. In BI-RADS, five types of margins are defined: circumscribed, obscured, microlobulated, ill-defined and spiculated. Usually ill-defined margins or spiculated lesions are much more likely to be malignant [7]. Density is a description of the overall breast compo-sition. In DDMS, the density value is between 1 and 4 where 1 means the breast is almost entirely fat, while 4 means the breast tissue is extremely dense [8].

The assessment descriptor in BI-RADS indicates the level of suspicion [8], which is a subjective interpretation and could vary among different radiologists. Since this descriptor cannot be computed or extracted from a mammogram image, the assess-ment descriptor was not used in our experiment.

In the experiment, we used four BI-RADS descriptors of masses: shape, margin, density, and patient age. The categories of the shape descriptor were converted into numeric values as: round = 1, oval = 2, lobulated = 3, irregular = 4 and architectural distortion = 5. Mass margin descriptors were also converted into numeric values as: circumscribed = 1, obscured =2, microlobulated = 3, ill-defined = 4 and spiculated= 5.

3 Methodology

3.1 Feature Selection

Feature selection is an important pre-processing step in classification. To obtain a good classification performance, it is critical to choose an optimal set of features for the given dataset [10]. To derive an optimal feature subset, we computed the Infor-mation Gain (IG) [11] for each descriptor. IG is a measure in Information Theory which indicates the informativeness of an attribute. Used in our context, IG essen-tially indicates the effectiveness of an attribute in the classification: a larger IG means the attribute is more informative. Gain(S,A) of an attribute A in a collection S is a measure based on Entropy:

$$Gain(S, A) = Entropy(S) - \sum_{v \in Values(A)} \frac{|S_v|}{|S|} Entropy(S_v)$$

where Value(A) is set of all possible values for attribute A, S_v is the subset of S with the attribute A of value v. And entropy [11] is a measure of purity, which can be used to indicate how pure a collection is. Entropy of a collection S is computed as:

$$Entropy(S) = -\sum_{j=1} p_j \log_2(p_j)$$

where p_j is the probability of the j subset in S (i.e., instances which belong to class j).

Note that the IG value is computed for each attribute separately, and the resulting values are not dependent on the order of attributes selected.

Figure 2 shows the Information Gain of each of the four BI-RADS descriptors we investigated. Of them, margin has the largest information gain, which indicates that mass margin is probably the most important descriptor/feature for classifying masses. On the other hand, density has much lower information gain, which indicates that the feature could be nearly irrelevant for classification.

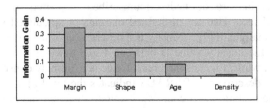

Fig. 2. Information Gain for each BI-RADS Descriptor

To verify this, we classified the dataset with and without the density descriptor and compared the performance with five different classifiers. Our experiment results (Table 1) show the classifications accuracies with density feature are slightly higher, however, the differences were not significantly for all five classifiers (p-value>0.05). Thus, we conclude that mass density descriptors were not important features, and can therefore be removed from the data without sacrificing classification performance.

Table 1. Classification Accuracies by Different BI-RADS Descriptors

BI-RADS Classifiers	Margin, Shape, Age, Density	Margin, Shape, Age	P-Value
Decision Tree	83.99 %	83.26 %	>0.05
Bayes Network	79.74 %	79.74 %	>0.05
Neural Network	84.88 %	84.73 %	>0.05
SVM	82.38 %	82.09 %	>0.05
KNN	86.78 %	84.43 %	>0.05

3.2 Converting Features to Coarse-Grained Categories

The original BI-RADS shape descriptor has five categories (round, oval, lobulated, irregular and architectural distortion). Our motivation for grouping them into coarse-grained categories was to simplify the decision process of classification. To this end, we first determined the splitting threshold by running the Decision Tree algorithm [11], where a decision tree was built using only the numeric shape feature with mass diagnosis as the target. Based on the decision tree, round, oval and lobulated shapes were categorized as "regular", and irregular and architectural distortion shapes were grouped as "irregular". In addition to the shape descriptor, we also converted the integer age feature into coarse-grained categories of "young" and "old". To determine the splitting threshold, we ran a Decision Tree and obtained the value of 57, in the same way as we did for the BI-RADS shape descriptor.

Note that, in our work in this paper, we converted the shape and age features to coarse-grained categories, but not the BI-RADS margin descriptor. That was because margin had by far the largest information gain over other features in our feature selection phase (as described earlier in section 3.1 above) – we assumed that finer categorization of this feature is critical in classifying masses.

3.3 Ensemble Learning and Partition Dataset into Subsets

Previous research has shown that ensemble learning often achieves better accuracy in classification than the individual classifiers that make them up [12]. In our work, the ensemble learning method partitions datasets and applies multiple classifiers as base classifications, and then combines the classifications from those partitioned datasets.

In our experiment, we tested three data partition schemes based on the features we converted to coarse-grained categories: 1) by age (young and old); 2) by shape (regular and irregular); 3) by age and shape (young regular, young irregular, old regular and old irregular). We expected that the overall classification accuracy could be improved by applying the best classification algorithms for each data subset. To find the best algorithms for a subset, we experimented with five classification algorithms: Decision Tree, Bayes Network, Neural Network, Support Vector Machine (SVM) and K-Nearest Neighbor (KNN). We chose those five algorithms because they have diverse characteristics. For example, Bayes Network is a statistical classifier; KNN decides the classification based on local information; Neural Network and SVM are known to be robust to noise. The final classification is the ensemble of multiple classifiers, where each classifier produces the highest accuracy among a data subset.

4 Experimental Results

First, we investigated the effect of converting the BI-RADS shape descriptor to coarse-grained categories. Table 2 below shows the accuracies by using the shape descriptor with coarse-grained categories, along with the margin descriptor and the age feature with original values. By comparing with the case using the original fine-grained shape categories, we can see that using the coarse-grained categories did not significantly decrease the accuracies for most classifiers (p-value > 0.05).

Table 2. Classification Accuracies for Fine vs. Coarse-grained Shape Categories

BI-RADS Classifier	Shape (fine-grained) Margin, Age	Shape (coarse-grained) Margin, Age	P-value
Decision Tree	83.26 %	83.26 %	>0.05
Bayes Network	79.74 %	79.74 %	>0.05
Neural Network	84.73 %	83.26%	>0.05
SVM	82.09 %	81.64 %	>0.05
KNN	84.43 %	80.32 %	**<0.05**

Next we investigated the ensemble learning for the shape descriptor. Table 3 below shows the classification accuracies of the regular vs. the irregular shape subsets. Since the coarse-grained categories of the shape descriptor were used to partition the dataset, only the BI-RADS margin descriptor and the patient age feature were used for classification. Column (a) "Weighted Accuracy" in the table indicates the average accuracies weighted by the proportion of the size of the subsets. Column (b) is the classification accuracies using the BI-RADS margin descriptor and the age feature without any dataset partition. With shape partition scheme, the classification accuracies for the partitioned datasets had no significant difference compared with the classifications without data partition for most classifiers (p-value < 0.05).

Table 3. Classification Accuracies with Partitioned Datasets by Shape vs. without Partition

Classifier	Regular Shape 454 instances	Irregular Shape 227 instances	Weighted Accuracy (a)	No Dataset Partition (b)	P-Value
Decision Tree	81.50 %	86.78 %	83.26 %	82.53 %	>0.05
Bayes Net	79.74 %	86.78 %	82.09 %	81.20 %	>0.05
Neural Network	80.62 %	86.78 %	82.67 %	81.64 %	>0.05
SVM	81.06 %	86.78 %	82.97 %	81.20 %	>0.05
KNN	77.53 %	85.46 %	80.18 %	76.21 %	**<0.05**
Best Classifier	81.50 % Decision Tree	86.78 % Decision Tree	83.54 %	82.53 % Decision Tree	>0.05

Then we investigated the effect of converting the age feature to coarse-grained categories. Table 4 shows the classification accuracies of the young vs. old subsets. All classifications included three features: the coarse-grained categories of the BI-RADS shape descriptor, the BI-RADS margin descriptor and the patient age. Column (a) "Weighted Accuracy" in this table is calculated in the same way as the previous table. Column (b) is the classification without data partition. For this partition scheme, the classification accuracies had no significant difference compared with the classifications without data partition for all classifiers (p-value < 0.05).

Finally we investigated the ensemble learning on four data subsets partitioned by age and shape. All classifications used only two features: the BI-RADS margin descriptor and the age feature. Table 5 shows the classification accuracies. Column (a) "Weighted Accuracy" is calculated in the same way as the previous tables. Column

Table 4. Classification Accuracies with Dataset Partition by Age vs. without Dataset Partition

Classifier	Young Age 348 instances	Old Age 333 instances	Weighted Accuracy (a)	No Dataset Partition (b)	P-Value
Decision Tree	85.63 %	81.08 %	83.41 %	83.26 %	>0.05
Bayes Net	81.90 %	81.68 %	81.79 %	79.74 %	>0.05
Neural Network	84.77 %	81.38 %	83.11 %	83.26 %	>0.05
SVM	81.61 %	79.88 %	80.76 %	81.64 %	>0.05
K NN	83.05 %	78.98 %	81.06 %	80.32 %	>0.05
Best Classifier	85.63 % Decision Tree	81.68 % Bayes Net	84.00 %	83.26 % Decision Tree	>0.05

Table 5. Classification Accuracies of Datasets Partitioned by Age and Shape

	Younger Age Regular Shape	Younger Age Irregular Shape	Older Age Regular Shape	Older Age Irregular Shape	Weighted Accuracy With Partition (a)	Accuracy Without Partition (b)	P-Value
Classifier	273 instances	75 instances	181 instances	152 instances	681 instances	681 instances	
Decision Tree	87.55 %	85.33 %	76.80 %	87.50 %	84.43%	82.53 %	>0.05
Bayes Net	85.35 %	85.33 %	72.93 %	87.50 %	82.53%	81.20 %	>0.05
Neural Network	84.62 %	82.67 %	75.69 %	87.50 %	82.67%	81.64 %	>0.05
SVM	82.42 %	85.33 %	76.24 %	87.50 %	82.23%	82.33%	>0.05
KNN	80.22 %	82.67 %	76.80 %	81.58 %	79.88%	76.21 %	**<0.05**
Best Classifiers	87.55 % (Decision Tree)	85.33 % (Decision Tree and others)	76.80 % (Decision Tree)	87.50 % (Decision Tree and others)	84.43% (*)	82.53 % Decision Tree	>0.05

* 84.43% is the weighted accuracy computed from the best classifiers of the last row.

(b) is the classification accuracies without dataset partition. The ensemble learning with weighted classification accuracy achieved better performance over the best classification with no data partitioning (84.43% vs. 82.53%). Note that, the weighted accuracy is largely dragged by the low accuracy from the older age and regular shape group, where three other groups have achieved significantly better classifications. To achieve significantly better classifications with the ensemble learning, we will further investigate and improve the classifications of the older age regular shape group in our future work.

5 Conclusions and Future Work

In this paper, we explored an ensemble learning of using quantized BI-RADS features for classifying masses in mammograms. Our experiment showed that mass density descriptor could be removed without sacrificing classification performance. Using the coarse-grained shape categories along with the margin descriptor and patient age, our ensemble classifier achieved the overall accuracy of 84.43%. Our results indicate that automatic clinical decision systems can be simplified by focusing on coarse-grained shape BI-RADS categories without losing any accuracy for classifying masses in mammograms.

In this experiment, we found that the mass instances of old age regular shape group produced much lower classification accuracies than other groups. This result suggests that the mass instances in this group are more difficult to classify, and using the age and the mass margin descriptor may not be enough to distinguish malignant from benign.

In future work, we are planning to apply the methods learned in this content-based classification system to build an image-based CADx system. And as our study indicates that margin is the most important feature for classifying masses, we plan to use even finer categorization such as continuous values to represent the margin feature in the image-based CADx system.

References

1. National Cancer Institute, American Cancer Society Cancer Facts & Figures (2008), http://www.cancer.org
2. Yankaskas, B.C., Schell, M.J., Bird, R.E., Desrochers, D.A.: Reassessment of Breast Cancers Missed During Routine Screening Mammography: A Community-Based. American Roentgen Ray Society 177, 535–541 (2001)
3. Cheng, H.D., Shi, X.J., Min, R., Hu, L.M., Cai, X.P., et al.: Approaches for Automated Detection and Classification of Masses in Mammograms. Pattern Recognition (2006)
4. D'Orsi, C.J., Bassett, L.W., Berg, W.A.: Breast Imaging Reporting and Data System: ACR BI-RADS-Mammography (ed 4). American College of Radiology, Reston, VA (2003)
5. Kim, S., Yoon, S.: BI-RADS Feature-Based Computer-Aided Diagnosis of Abnormalities in Mammographic. In: 6th International Special Topic Conference on ITAB (2007)
6. Elter, M., Schulz-Wendtland, R., Wittenberg, T.: Prediction of Breast Biopsy Outcomes Using CAD Approaches That Both Emphasize an Intelligible Decision Process. Medical Physics 34(11) (2007)
7. Winchester, D.J., Winchester, D.P., Hudis, C.A., Norton, L.: Breast Cancer, 2nd edn. Springer, Heidelberg (2007)
8. The Digital Database for Screening Mammography, http://marathon.csee.usf.edu/Mammography/Database.html
9. Witten, I., Frank, E.: Data Mining: Practical Machine Learning Tools and Techniques, 2nd edn. Morgan Kaufmann, San Francisco (2005)
10. Kohavi, R., John, G.: Wrappers for Feature Subset Selection. Artificial Intelligence 97(1-2), 273–324 (1997)
11. Mitchell, T.M.: Machine Learning. McGraw-Hill, New York (2001)
12. Opitz, D., Maclin, R.: Popular Ensemble Methods: An Empirical Study. Journal of Artificial Intelligence Research 11, 169–198 (1999)

Robust Learning-Based Annotation of Medical Radiographs

Yimo Tao[1,2], Zhigang Peng[1], Bing Jian[1], Jianhua Xuan[2], Arun Krishnan[1],
and Xiang Sean Zhou[1]

[1] CAD R&D, Siemens Healthcare, Malvern, PA USA
[2] Dept. of Electrical and Computer Engineering, Virginia Tech, Arlington, VA USA

Abstract. In this paper, we propose a learning-based algorithm for automatic medical image annotation based on *sparse aggregation of learned local appearance cues*, achieving high accuracy and robustness against severe diseases, imaging artifacts, occlusion, or missing data. The algorithm starts with a number of landmark detectors to collect local appearance cues throughout the image, which are subsequently verified by a group of learned *sparse spatial configuration models*. In most cases, a decision could already be made at this stage by simply aggregating the verified detections. For the remaining cases, an additional global appearance filtering step is employed to provide complementary information to make the final decision. This approach is evaluated on a large-scale chest radiograph view identification task, demonstrating an almost perfect performance of 99.98% for a posteroanterior/anteroposterior (PA-AP) and lateral view position identification task, compared with the recently reported large-scale result of only 98.2% [1]. Our approach also achieved the best accuracies for a three-class and a multi-class radiograph annotation task, when compared with other state of the art algorithms. Our algorithm has been integrated into an advanced image visualization workstation, enabling content-sensitive hanging-protocols and auto-invocation of a computer aided detection algorithm for PA-AP chest images.

1 Introduction

The amount of medical image data produced nowadays is constantly growing, and a fully automatic image content annotation algorithm can significantly improve the image reading workflow, by automatic configuration/optimization of image display protocols, and by off-line invocation of image processing (e.g., denoising or organ segmentation) or computer aided detection (CAD) algorithms. However, such annotation algorithm must perform its tasks in a *very accurate* and *robust* manner, because even "occasional" mistakes can shatter users' confidence in the system, thus reducing its usability in the clinical settings. In the radiographic exam routine, chest radiograph comprise at least one-third of all diagnostic radiographic procedures. Chest radiograph provides sufficient pathological information about cardiac size, pneumonia-shadow, and mass-lesions, with low cost and high reproducibility. However, about 30%-40% of the projection

B. Caputo et al. (Eds.): MCBR-CDS 2009, LNCS 5853, pp. 77–88, 2010.
© Springer-Verlag Berlin Heidelberg 2010

and orientation information of images in the DICOM header are unknown or mislabeled in the picture archive and communication system (PACS) [2]. Given a large number of radiographs to review, the accumulated time and cost can be substantial for manually identifying the projection view and correcting the image orientation for each radiograph.

The goal of this study is to develop a *highly accurate* and *robust* algorithm for automatic annotation of medical radiographs based on the image data, correcting potential errors or missing tags in the DICOM header. Our first focus is to automatically recognize the projection view of chest radiographs into posteroanterior/anteroposterior (PA-AP) and lateral (LAT) views. Such classification could be exploited on a PACS workstation to support optimized image hanging-protocols [1]. Furthermore, if a chest X-ray CAD algorithm is available, it can be invoked automatically on the appropriate view(s), saving users' *manual effort* to invoke such an algorithm and the potential *idle time* while waiting for the CAD outputs. We also demonstrate the algorithm's capability of annotating other radiographs beyond chest X-ray images, in a three-class setting and a multi-class setting. In both cases, our algorithm significantly outperformed existing methods.

A great challenge for automatic medical image annotation is the large visual variability across patients in medical images from the same anatomy category. The variability caused by individual body conditions, patient ages, and diseases or artifacts would fail many seemingly plausible heuristics or methods based on global or local image content descriptors. Fig. 1 and Fig. 2 show some examples of PA-AP and LAT chest radiographs. Because of obliquity, tilt, differences in projection, and the degree of lung inflation, the same class PA-AP and LAT images may present very high inter patient variability. Fig. 3 shows another example of images from the "pelvis" class with considerable visual variation caused by differences in contrast, field of view (FoV), diseases/implants, and imaging artifacts.

Most existing methods (e.g., [3], [4]) for automatic medical image annotation were based on different types of image content descriptors, separately or combined together with different classifiers. Müller et al. [5] proposed a method using weighted combinations of different global and local features to compute the similarity scores between the query image and the reference images in the training database. The annotation strategy was based on the GNU Image Finding Tool image retrieval engine. Güld and Deserno [6] extracted pixel intensities from down-scaled images and other texture features as the image content descriptor. Different distance measures were computed and summed up in a weighted combination form as the final similarity measurement used by the nearest-neighbor decision rule (1NN). Deselaers and Ney [4] used a bag-of-features approach based on local image descriptors. The histograms generated using bags of local image features were classified using discriminative classifiers, such as support vector machine (SVM) or 1NN. Keysers et al. [7] used a nonlinear model considering local image deformations to compare images. The deformation measurement was then used to classify the image using 1NN. Tommasi et al. [8] extracted SIFT [9]

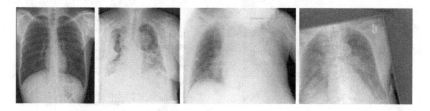

Fig. 1. The PA-AP chest images of normal patient, patients with severe chest disease, and an image with unexposed region on the boundary

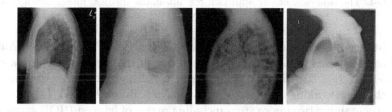

Fig. 2. The LAT chest images of normal patient, patients with severe chest disease, and an image with body rotation

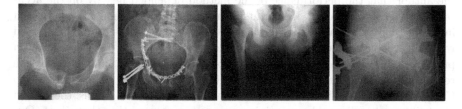

Fig. 3. Images from the IRMA/ImageCLEF2008 database with the IRMA code annotated as: acquisition modality "overview image"; body orientation "AP unspecified"; body part "pelvis"; biological system "musculoskeletal" . Note the very high appearance variability caused by artifacts, diseases/implants, and different FoVs.

features from downscaled images and used the similar bag-of-features approach [4]. A modified SVM integrating the bag-of-features and pixel intensity features was used for classification.

Regarding the task for recognizing the projection view of chest radiographs, Pieka and Huang [10] proposed a method using two projection profiles of images. Kao et al. [11] proposed a method using a linear discriminant classifier (LDA) with two features extracted from horizontal axis projection profile. Aimura et al. [12] proposed a method by computing the cross-correlation coefficient based similarity of an image with manually defined template images. Although high accuracy was reported, manually generation of those template images from a large training image database was time consuming and highly observer dependent. Lehman et al. [13] proposed a method using down-scaled image pixels with

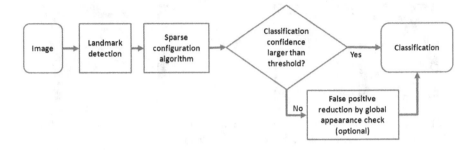

Fig. 4. The overview of our approach for automatic medical image annotation

four distance measures along with K-nearest neighbor (KNN) classifier. Almost equal accuracy was reported when compared with the method of Aimura et al. [12] on their test set. Boone [2] developed a method using a neural network (NN) classifier working on down-sampled images. Recently, Luo [1] proposed a method containing two major steps including region of interest (ROI) extraction, and then classification by the combination of a Gaussian mixture model classifier and a NN classifier using features extracted from ROI. An accuracy of 98.2% was reported on a large test set of 3100 images. However, it was pointed out by the author that the performance of the method depended heavily on the accuracy of ROIs segmentation. Inaccurate or inconsistent ROI segmentations would introduce confusing factors to the classification stage. All the aforementioned work regarded the chest view identification task as a two class classification problem, however, we included an additional OTHER class in this work. The reason is that in order to build a fully automatic system to be integrated into CAD/PACS for identification of PA-AP and LAT chest radiographs, the system must filter out radiographs containing anatomy contents other than chest. Our task, therefore, becomes a three-class classification problem, i.e., identifying images of PA-AP, LAT, and OTHER, where "OTHER" are radiographs of head, pelvis, hand, spine, etc.

In this work, we adopt a hybrid approach based on *robust aggregation of learned local appearance findings*, followed by the exemplar-based global appearance filtering. Fig. 4 shows the overview of the proposed algorithm. Our algorithm is designed to first detect multiple focal anatomical structures within the medical image. This is achieved through a learning-by-example landmark detection algorithm that performs simultaneous feature selection and classification at several scales. A second step is performed to eliminate inconsistent findings through a robust *sparse spatial configuration* (SSC) algorithm, by which consistent and reliable local detections will be retained while outliers will be removed. Finally, a reasoning module assessing the fitered findings, i.e., remaining landmarks, is used to determine the final content/orientation of the image. Depending on the classification task, a post-filtering component using the exemplar-based global appearance check for cases with low classification confidence may also be included to reduce false positive (FP) identifications.

2   Methods

2.1   Landmark Detection

Anatomical landmark detection plays a fundamental and critical role for medical image analysis. High level medical image understanding usually starts from the identification and localization of anatomical structures. Therefore, accurate anatomical landmark detection becomes critical.

The landmark detection module in this work was inspired by the work of Viola and Jones [14], but modified to detect points (e.g., the carina of trachea) instead of a fixed region of interest (e.g., a face). We use an adaptive coarse-to-fine implementation in the scale space, and allow for flexible handling of the *effective scale* of anatomical context for each landmark. More specifically, we train landmark detectors independently at several scales. For this application, two scales are sufficient to balance the computational time and detection accuracy. During the training phase, for a landmark at a specific scale, a sub-patch that covers the sufficient and effective context of an anatomy landmark is extracted; then an over-complete set of extended Haar features are computed within the patch. In this work, the size of the sub-patches for each landmark varies from 13×13 to 25×25 pixels depending on its position in the image. The sub-patches are allowed to extend beyond the image border, in which case the part of the patch falling outside the image is padded with zeroes. For classification, we employ the boosting framework [15] for simultaneous feature selection and classification.

During the testing/detection phase, the trained landmark detectors at the coarsest scale are used first to scan on the whole image to determine the candidate position(s), where the response(s)/detection score(s) are larger than the predefined threshold. After that, the landmark detectors at finer scales are scrutinized at previously determined position(s) to locate the local structures more accurately and, thus, to obtain the final detection. The final outputs of a landmark detector are the horizontal and vertical (x-y) coordinates in the image along with a response/detection score. Joint detection of multiple landmarks also proves beneficial (see Zhan et al.[16] for detail).

2.2   Reasoning Strategy

Knowing that the possible locations of anatomical landmarks are rather limited, we aim to exploit this geometric property to eliminate the possible redundant and erroneous detections from the first step. This geometric property is represented by a spatial constellation model among detected landmarks. The evaluation of consistency between a landmark and the model can be determined by the spatial relationship between the landmark and other landmarks, i.e., how consistent the landmark is according to other landmarks. In this work, we propose a local voting algorithm (Alg. 1) to sequentially remove false detections until no outliers exist. The main idea is that each detected landmark is considered as a candidate and the quality of such candidate is voted upon by voting groups formed by other landmarks. A higher vote means the candidate is more likely to be a good local feature.

Algorithm 1. Sparse spatial configuration algorithm

---

for each candidate x_i **do**
  for each combinations of $X \backslash x_i$ **do**
    Compute the vote of x_i
  end for
  Sort all the votes received by landmark x_i. (The sorted array is defined by γ_{x_i}).
end for
repeat
  $\check{x} = \arg\min_{x_i} \max \gamma_{x_i}$
  Remove $\check{x}$ and all votes involved with $\check{x}$.
until Only M candidates are left

In general, our reasoning strategy "peels away" erroneous detections in a sequential manner. Each candidate x receives a set of votes from other candidates. We denote the ith detected landmark as x_i, which is a two dimensional variable with values corresponding to the detected x-y coordinates in the image. The vote received by candidate x_i is denoted by $\eta(x_i|X_\nu)$, where X_ν is a voting group containing other landmarks. The vote is defined as the likelihood between candidate x_i and its predicted position ν_i coming from the voting group. The likelihood function is modeled as multi-variant Gaussian as following:

$$\eta(x_i|X_\nu) = \frac{1}{2\pi |\Sigma|^{1/2}} e^{-(x_i-\nu_i)^T \Sigma^{-1}(x_i-\nu_i)} \tag{1}$$

where Σ is the estimated covariance matrix, and the prediction $\nu_i = q(x_i|X_\nu)$. Here $q(\bullet)$ is defined as:

$$q(x_i|X_\nu) = A \times [X_\nu] \tag{2}$$

where A is the transformation matrix learned by linear regression from a training set, and $[X_\nu]$ is the array formed by the x-y coordinates of landmarks from the voting group X_ν. The voting groups for x_i are generated by the combinations of several landmarks from the landmark set excluding x_i (denoted as $X \backslash x_i$). The size of each voting group is designed to be small, so that the resultant *sparse* nature guarantees that the shape prior constraint could still take effect even with many missed detections, thus leading its robustness in handling challenging cases such as those with a large percentage of occlusion, or missing data. In this work, we set the sizes of the voting groups to be 1 to 3.

The reasoning strategy (Alg. 1) then iteratively determines whether to remove the current "worst" candidate, which is the one with the smallest maximum vote score compared with other candidates. The algorithm will remove the "worst" candidate if its vote score is smaller than a predefined vote threshold $V_{threshold}$. This process will continue until no landmark outlier exists. The bad candidates can be effectively removed by this strategy.

2.3   Classification Logic

The classification logic using the filtered landmarks is straightforward. The number of remaining landmarks for each image class is divided by the total number of detectors for that class, representing the final classification score. In case that equal classification scores are obtained between several classes, the average landmark detection scores are used to choose the final single class. Depending on the classification task, a FP reduction module based on the global appearance check may also be used for those images with low classification confidence. A large portion of these images come from the OTHER class. They have a small number of local detections belonging to the candidate image class, yet their spatial configuration is strong enough to pass the SSC stage. Since the mechanism of local detection integration from previous steps could not provide sufficient discriminative information for classification, we try to integrate a post-filtering component based on the global appearance check to make the final decision. In our experiment for PA-AP/LAT/OTHER separation task, only about 6% of cases go through this stage. To meet the requirement for real-time recognition, an efficient exemplar-based global appearance check method is adopted. Specifically, we use pixel intensities from 16×16 down-scaled image as the feature vector along with 1NN, which uses the Euclidean distance as the similarity measurement. With the fused complementary global appearance information, the FP reduction module could effectively remove FP identified images from the OTHER class, thus leading to the overall performance improvement of the final system (see Section 3).

3   Results

3.1   Datasets

We ran our approach on four tasks: PA-AP/LAT chest radiograph view position identification with and without OTHER class using a large-scale in house database, and the multi-class medical radiograph annotation with and without OTHER class using the ImageCLEF2008 database [1].

1) The in-house image database were collected from daily imaging routine from radiology departments in hospitals, containing a total of 10859 radiographs including 5859 chest radiographs and 5000 other radiographs from a variety of other anatomy classes. The chest images covered a large variety of chest exams, representing image characteristics from real world PACS. We randomly selected 500 PA-AP, 500 LAT, and 500 OTHER images for training landmark detectors. And the remaining images are used as testing set.

2) For the multi-class medical radiograph annotation task, we selected the top nine classes which have the most number of images from the ImageCLEF2008 database. The selected nine classes included PA-AP chest, LAT chest, PA-AP left hand, PA-AP cranium, PA-AP lumbar spine, PA-AP pelvis, LAT lumbar

---

[1] http://imageclef.org/2008/medaat

spine, PA-AP cervical spine, and LAT left to right cranium. The remaining images were regarded as one OTHER class. We directly used the chest landmark detectors from the previous task. 50 PA-AP and 50 LAT chest testing images were randomly seleted from the testing set of previous task. For the remaining 7 classes, we randomly selected 200 (150 training / 50 testing) images for each class. For OTHER class, we used 2000 training and 2000 testing images each. All images were down-scaled to have the longest edge of 512 pixels while preserving the aspect ratio.

3.2   Classification Precision

For the chest radiograph annotation task, we compared our method with three other methods described by Boone et al. [2], Lehmann et al. [13], and Kao et al. [11]. For method proposed by Boone et al. [2], we down-sampled the image to the resolution of 16×16 pixels and constructed a five hidden nodes NN. For method proposed by Lehmann et al. [13], a five nearest neighbor (5-NN) classifier using 32×32 down-sampled image with the correlation coefficient distance measurement was used. The same landmark detector training database was used as the reference database for the 5-NN classifier. For method proposed by Kao et al. [11], we found that the projection profile derived features described in the literature were sensitive to the orientation of anatomy and noise in the image. Directly using the smoothed projection profile as the feature along with the LDA classifier provided better performance. Therefore, we used this improved method as our comparison.

For the multi-class radiograph annotation task, we compared our method with the in-house implemented bag-of-features method proposed by Deselaers and Ney [4] (named as PatchBOW+SVM) and the method proposed by Tommasi et al. [8] (named as SIFTBOW+SVM). Regarding PatchBOW+SVM, we used the bag-of-features approach based on randomly cropped image sub-patches. The generated bag-of-features histogram for each image had 2000 bins, which were then classified using a SVM classifier with a linear kernel. Regarding SIFT-BOW+SVM, we implemented the same modified version of SIFT (modSIFT) descriptor and used the same parameters for extracting bag-of-features as those used by Tommasi et al. [8]. We combined the 32×32 pixel intensity features and the modSIFT bag-of-features as the final feature vector, and we used a SVM classifier with a linear kernel for classification. We also tested the benchmark performance of directly using 32×32 pixel intensity from the down-sampled image as the feature vector along with a SVM classifier.

Table 1 and 2 show the performance of our method along with other methods. It is seen that our system has obtained almost perfect performance on the PA-AP/LAT separation task. The only one failed case is a pediatric PA-AP image. Our method also performed the best on the other three tasks. Fig. 5 shows the classification result along with the detected landmarks for different classes. It can be seen that our method could robustly recognize challenging cases under the influence of artifacts or diseases.

Table 1. PA-AP/LAT/OTHER chest radiographs annotation performance

	PA-AP/LAT	PA-AP/LAT/ OTHER
Our method	-	**98.81%**
Our method without FP reduction	**99.98%**	98.47%
Lehmann's method	99.04%	96.18%
Boone's method	98.24%	-
Improved Projection Profile method	97.60%	-

Table 2. Multi-class radiographs annotation performance

	Mutli-class without OTHER	Multi-class with OTHER
Our method	**99.33%**	**98.81%**
Subimage pixel intensity + SVM	97.33%	89.00%
PatchBOW + SVM	96.89%	94.71%
SIFTBOW + SVM	98.89%	95.86%

3.3 Intermediate Results

Landmark Detection: We provide here the intermediate results of landmark detectors' performance. In this work, we used 11 landmarks and 12 landmarks for PA-AP and LAT chest images. As for the multi-class radiograph annotation task, we used 7-9 landmarks for other image classes. The selection of landmarks was according to Netter [17]. To test the landmark detectors' performance, we annotated 100 PA-AP and 100 LAT images separately. Since the landmark detectors run on the Gaussian smoothed images, the detected position could deviate from the ground truth position to certain degree, which is allowable for our image annotation application. We determine the detected landmark as true positive detection when the distance between the detected position and the annotated ground truth position is smaller than 30 pixels. Note that the detection performance can be traded off against computational time. Currently in order to achieve real-time performance, we accepted an average sensitivity for the 23 chest landmark detectors at 86.91% (±9.29%), which was good enough to support the aforementioned overall system performance.

SSC: For the PA-AP/LAT separation task on the 200 images where ground truth landmarks were annotated, 55 out of 356 false positive landmark detections were filtered by the SSC algorithm, while the true positive detections were unaffected. In addition, the algorithm removed 921 and 475 false positive detections for the PA-AP/LAT/OTHER task and the multi-class task with OTHER class. Fig. 6 shows that the result of the voting algorithm in reducing false positive detections on non-chest image classes. We can conclude that the voting strategy has improved the specificity of the landmark detectors.

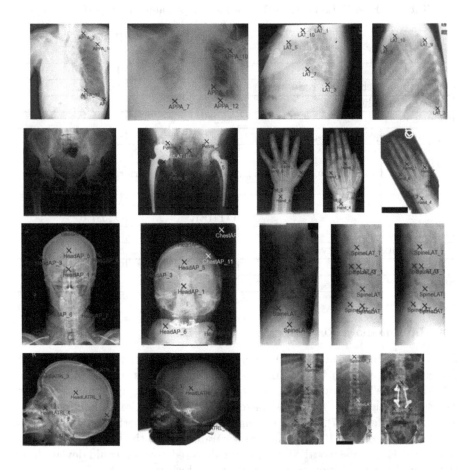

Fig. 5. Examples of the detected landmarks on different images

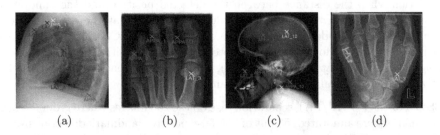

(a)   (b)   (c)   (d)

Fig. 6. The SSC algorithm performance on different image classes (better viewed in color): (a) LAT chest, (b) foot, (c) cranium, and (d) hand. The blue colored crosses are true positive landmark detections; the yellow colored ones are false positive detections; and the red colored ones are detections filtered by the SSC algorithm. APPA and LAT label under the detected landmarks specify that detections are from PA-AP chest detectors or LAT chest detectors.

4   Conclusion

To conclude, we have developed a hybrid learning-based approach for parsing and annotation of medical radiographs. Our approach integrates learning-based local appearance detections, the shape prior constraint by a sparse configuration algorithm, and a final filtering stage with the exemplar-based global appearance check. This approach is highly accurate, robust, and fast in identifying images even when altered by diseases, implants, or imaging artifacts. The robustness and efficiency of the algorithm come from: (1) the accurate and fast local appearance detection mechanism with the sparse shape prior constraint, and (2) the complementarity of local appearance detections and the global appearance check. The experimental results on a large-scale chest radiograph view position identification task and a multi-class medical radiograph annotation task have demonstrated the effectiveness and efficiency of our method. As a result, minimum manual intervention is required, improving the usability of such systems in the clinical environment. Our algorithm has already been integrated into an advanced image visualization workstation for enabling content-sensitive hanging-protocols and auto-invocation of a CAD algorithm on identified PA-AP chest images. Due to the generality and scalability of our approach, it has the potential to annotate more image classes from other categories and modalities.

References

1. Luo, H., Hao, W., Foos, D., Cornelius, C.: Automatic image hanging protocol for chest radiographs in PACS. IEEE Transactions on Information Technology in Biomedicine 10(2), 302–311 (2006)
2. Boone, J.M., Hurlock, G.S., Seibert, J.A., Kennedy, R.L.: Automated recognition of lateral from PA chest radiographs: saving seconds in a PACS environment. Journal of Digital Imaging 16(4), 345–349 (2003)
3. Deselaers, T., Deserno, T.M., Müller, H.: Automatic medical image annotation in ImageCLEF 2007: overview, results, and discussion. Pattern Recognition Letters 29(15), 1988–1995 (2008)
4. Deselaers, T., Ney, H.: Deformations, patches, and discriminative models for automatic annotation of medical radiographs. Pattern Recognition Letters 29(15), 2003–2010 (2008)
5. Müller, H., Gass, T., Geissbuhler, A.: Performing image classification with a frequency-based information retrieval schema for ImageCLEF 2006. In: Image-CLEF 2006. working notes of the Cross Language Evalutation Forum (CLEF 2006), Alicante, Spain (2006)
6. Güld, M.O., Deserno, T.M.: Baseline results for the ImageCLEF 2007 medical automatic annotation task using global image features. In: Peters, C., Clough, P., Gey, F.C., Karlgren, J., Magnini, B., Oard, D.W., de Rijke, M., Stempfhuber, M. (eds.) CLEF 2006. LNCS, vol. 4730, pp. 637–640. Springer, Heidelberg (2007)
7. Keysers, D., Deselaers, T., Gollan, C., Ney, H.: Deformation models for image recognition. IEEE Transactions on Pattern Analysis and Machine Intelligence 29(8), 1422–1435 (2007)
8. Tommasi, T., Orabona, F., Caputo, B.: Discriminative cue integration for medical image annotation. Pattern Recognition Letters 29(15), 1996–2002 (2008)

9. Lowe, D.: Distinctive image features from scale-invariant keypoints. International Journal on Computer Vision 60(2), 91–110 (2004)
10. Pietka, E., Huang, H.K.: Orientation correction for chest images. Journal of Digital Imaging 5(3), 185–189 (1992)
11. Kao, E., Lee, C., Jaw, T., Hsu, J., Liu, G.: Projection profile analysis for identifying different views of chest radiographs. Academic Radiology 13(4), 518–525 (2006)
12. Arimura, H., Katsuragawa, S., Ishida, T., Oda, N., Nakata, H., Doi, K.: Performance evaluation of an advanced method for automated identification of view positions of chest radiographs by use of a large database. In: Proceeding of SPIE Medical Imaging, vol. 4684, pp. 308–315 (2002)
13. Lehmann, T.M., Güld, O., Keysers, D., Schubert, H., Kohnen, M., Wein, B.B.: Determining the view of chest radiographs. Journal of Digital Imaging 16(3), 280–291 (2003)
14. Viola, P., Jones, M.: Rapid object detection using a boosted cascade of simple features. In: IEEE Conference on Computer Vision and Pattern Recognition (CVPR 2001), vol. 1, pp. 511–518 (2001)
15. Freund, Y., Schapire, R.: A decision-theoretic generalization of on-line learning and an application to boosting. Journal of computer and system sciences 55(1), 119–139 (1997)
16. Zhan, Y., Zhou, X.S., Peng, Z., Krishnan, A.: Active scheduling of organ detection and segmentation in whole-body medical images. In: Metaxas, D., Axel, L., Fichtinger, G., Székely, G. (eds.) MICCAI 2008, Part I. LNCS, vol. 5241, pp. 313–321. Springer, Heidelberg (2008)
17. Netter, F.H.: Atlas of human anatomy, 4th edn. Netter Basic Science. Elsevier Health Sciences (2006)

Knowledge-Based Discrimination in Alzheimer's Disease

Simon Duchesne[1,2], Burt Crépeault[1], and Carol Hudon[1,3]

[1] Centre de Recherche Université Laval – Robert Giffard, Québec, QC, Canada
[2] Département de radiologie, Université Laval, Québec, QC, Canada
[3] École de Psychologie, Université Laval, Québec, QC, Canada

Abstract. Our goal is to propose a methodology to integrate clinical, cognitive, genetic and neuroimaging results, a disparate assembly of categorical, ordinal, and numerical data, within the context of Alzheimer's disease (AD) research. We describe a knowledge-based, explicit decision model for the discrimination of AD subjects from normal controls (CTRL) based on a recent approach for diagnostic criteria for AD. We proceed by (a) creating a set of rules for each datum source; (b) integrating these disparate data into an information feature in the form of a binary string; and (c) using a machine learning approach based on the Hamming distance as an information similarity measure for the discrimination of probable AD from CTRL. When tested on 249 subjects from the multicentric Alzheimer's Disease Neuroimaging Initiative study, we reach up to 99.8% discriminative accuracy using a 9-bit information string in a series of 10-fold validation experiments. We expect this versatile approach to become useful in the identification of specific and sensitive phenotypes from large amounts of disparate data.

Keywords: Alzheimer's disease, knowledge-based model, multi-source data, discrimination.

1 Introduction

Early detection of Alzheimer's disease (**AD**) is critical for treatment success and constitutes a high priority research area. Pathologically-confirmed diagnostic accuracy of baseline clinical testing for AD averages **78%** (22% error rate) [1, 2] with insufficient diagnostic specificity [3]. To increase the accuracy of this procedure, recent thinking on clinical diagnostic criteria for AD stresses the importance of relying on a *combination* of core (clinical / cognitive) and supportive (neuroimaging / genetic / proteomic) assessment techniques [3].

This article describes our work towards a *knowledge-based methodology* for the integration of clinical, cognitive, genetic and neuroimaging results, a disparate assembly of categorical, ordinal, and numerical data, within the context of AD research. The general goal in combining information from these different modalities is to enable machine learning – based classification that may achieve a more accurate discrimination between normal aging and probable AD than *either of these data sources can provide alone*.

While these measures are routinely correlated with each other (e.g. in regression models), few knowledge-based systems exist in the context of probable AD discrimination,

B. Caputo et al. (Eds.): MCBR-CDS 2009, LNCS 5853, pp. 89–96, 2010.
© Springer-Verlag Berlin Heidelberg 2010

and those are typically focused on a single-modality [4]. Our proposed technique integrates disparate categorical, ordinal and numerical information from multi-source data, based on an *a priori* and explicit discriminative model. We aim to extract phenotypes related to either probable AD and CTRL subjects, and hypothesize that these phenotypes can be used in a machine learning setting to accurately discriminate between CTRL and probable AD. Our main intent with this article is to *propose* an integrative methodology that can be extended and applied to other pathologies, rather than *find* the optimal knowledge function for the identification of probable AD and its discrimination from normal aging.

2 Methods

2.1 General Approach

We propose a knowledge-based, algorithmic approach making explicit the decision model of the proposed new research criteria for the diagnosis of AD recently reported in Dubois et al. [3]. In this new approach, an affected individual must fulfill the core clinical criterion and at least one or more of the supportive biomarker criteria to meet the criteria for probable AD. For purposes of generality, we add model risk factors known to influence and moderate the risk of developing AD:

(A) *Core diagnostic criterion: presence of early episodic memory impairment.* Neuropsychological test results can attest to the presence of an early and significant episodic memory impairment as objective evidence of significantly impaired episodic memory on testing. We propose a *composite score* based on relevant information from a variety of cognitive tests to measure this deficit in episodic memory;

(B) *Supportive diagnostic criterion: presence of medial temporal lobe atrophy.* Quantitative volumetry in regions of interest such as the hippocampi and entorhinal cortex measures the loss related to the disease, when referenced to a well-characterized population. To this end we include normalized, quantitative hippocampal volumes obtained with minimal bias via semi-automatic extraction from MRI; and

(C) *Risk factors: age, gender, education, and APOE genotype.* Age and APOE are respectively known as the major demographic and genetic risk factors for the development of AD [5]. Authors have proposed that gender-specificity of neuropsychological performance needs to be accounted for in clinical diagnosis of AD [6]. Education may modulate the degree to which the neuropathology of early AD is expressed as impaired cognitive performance [7].

We postulate that the probability of an individual to be diagnosed with probable AD is therefore a function combining the core, supportive, and risk criteria:

$$p(probable\ AD) = p(core;\ supportive;\ risk)$$

We created a set of rules for each datum source (e.g. z-score levels) of the preceding criteria. This approach allowed us to integrate these disparate data into an information feature in the form of a binary string, allowing the use of a machine learning approach based on the Hamming distance [8] as an information similarity measure, for the purpose of discriminating probable AD from CTRL.

2.2 Subjects

Data used in the preparation of this article were obtained from the Alzheimer's Disease Neuroimaging Initiative (ADNI) database. The initial goal of ADNI was to recruit 800 adults, ages 55 to 90, to participate in the research -- approximately 200 cognitively normal older individuals to be followed for 3 years, 400 people with MCI to be followed for 3 years, and 200 people with early AD to be followed for 2 years, from over 50 sites across the U.S. and Canada. For up-to-date information see www.adni-info.org. Out of a final, total baseline recruitment of 392 CTRL and probable AD, we selected 249 subjects (138 CTRL; 111 AD) based on their (a) having consistent baseline to 6-months follow-up clinical diagnostic; and (b) for which the ensemble of clinical, cognitive, genomic, and hippocampal volumetry results were available at the time of submission.

2.3 Core Criterion: Episodic Memory Impairment Composite Score

Each site in the ADNI study complied in providing a Uniform Data Set for each subject enrolled, which includes 12 cognitive tests that are widely used in multi-center trials studying CTRL and early AD subjects. We focused on three specific tests, namely the Mini-Mental State Examination (MMSE) [9], the Logical Memory Test (LMT), and the Alzheimer's Disease Assessment Scale – Cognitive (ADAS-Cog)[10], as the tests that include components which specifically assess episodic memory. We propose a *composite score* based on these three tests to objectively attest to the presence of a deficit in episodic memory.

The MMSE is a fully structured screening instrument frequently used for AD drug studies. We selected only the three delayed word recall elements. For each individual i we computed a score $MMSE_i$ as the summation of results in these three questions (best score: 6; worse score: 0). We then proceeded into turning the individual results into z-scores, based on the distribution of MMSE score results from the overall CTRL population. We transformed the z-scores into a binary unit given the following rule:

$$z_i^{MMSE} = \frac{MMSE_i - \overline{MMSE}_{CTRL}}{\sigma_{MMSE_{CTRL}}} \; ; \; b_{MMSE_i} = \begin{cases} 1 & if(z_i^{MMSE} \leq -1.65) \\ 0 & otherwise \end{cases} \quad \textbf{Eq. (1, 2)}$$

whereby the 1.65 z-score threshold corresponds to a 5% rejection region in a normal distribution.

The ADAS cog is a structured scaled with some elements evaluating episodic memory via immediate word recall, delayed word recall, and word recognition. For each individual i we computed a score $ADAS_i$ as the summation of results in these three questions (best score: 0; worse score: 30), and proceeded with similar transformations to arrive at the bit unit:

$$z_i^{ADAS} = \frac{ADAS_i - \overline{ADAS}_{CTRL}}{\sigma_{ADAS_{CTRL}}} \; ; \; b_{ADAS_i} = \begin{cases} 1 & if(z_i^{ADAS} \geq -1.65) \\ 0 & otherwise \end{cases} \quad \textbf{Eq. (3, 4)}$$

The LMT is a modification of the episodic memory measure from the Wechsler Logical Memory – Revised (**WLM-R**)[11]. In this modified version, free recall of one short story is elicited immediately after reading and again after a thirty minutes delay. The total bits of information from the story that are recalled immediately (best score = 25) and after the delay interval (best score = 25) are recorded. We computed a total score (best score = 50) by adding both results into a single LMT_i score, which we turned into a binary unit in a similar process as above:

$$z_i^{LMT} = \frac{LMT_i - \overline{LMT}_{CTRL}}{\sigma_{LMT_{CTRL}}} \; ; \; b_{LMT_i} = \begin{cases} 1 & if(z_i^{LMT} \le -1.65) \\ 0 & otherwise \end{cases} \qquad \textbf{Eq. (5 , 6)}$$

2.4 Supportive Criterion: Hippocampal Volumetry

The protocol selected for the ADNI study included a 3D magnetization prepared rapid gradient echo (MP-RAGE) scan [12]. From the baseline scans a semi-automated technique was used to segment both left and right hippocampal volumes [13]. For each individual i we transformed the left/right hippocampal volumes (LHC_i, RHC_i) to arrive at individual bit units (e.g.):

$$z_i^{LHC} = \frac{LHC_i - \overline{LHC}_{CTRL}}{\sigma_{LHC_{CTRL}}} \; ; \; b_{LHC_i} = \begin{cases} 1 & if(z_i^{LHC} \le -1.65) \\ 0 & otherwise \end{cases} \qquad \textbf{Eq. (8 , 9)}$$

2.5 Risk Factors

Baseline age and years of education were transformed into z-scores and bit units in a similar fashion as described above:

$$z_i^{age} = \frac{age_i - \overline{age}_{CTRL}}{\sigma_{age_{CTRL}}} \; ; \; b_{AGE_i} = \begin{cases} 1 & if(z_i^{age} \le -1.65) \\ 0 & otherwise \end{cases} \; ; \; \textbf{Eq. (10,11)}$$

$$z_i^{scol} = \frac{scol_i - \overline{scol}_{CTRL}}{\sigma_{scol_{CTRL}}} \; ; \; b_{SCOL_i} = \begin{cases} 1 & if(z_i^{scol} \le -1.65) \\ 0 & otherwise \end{cases} \; ; \; \textbf{Eq. (12, 13)}$$

Gender was transformed into a binary unit if the subject's gender $GENDER_i$ matched the most occurring gender in the probable AD populations:

$$b_{GENDER_i} = \begin{cases} 1 & if(GENDER_i = \max(prob(gender_{AD})) \\ 0 & otherwise \end{cases} \qquad \textbf{Eq. (14)}$$

ADNI subjects were classified on a multiplicative risk scale according to the APOE genotyping results for allele 1 and 2 by multiplying the number of repetitions for both alleles and setting a risk threshold, such that subjects with so-called 'protective' or moderate risk alleles (two 2 alleles; one 2 allele and one 3 allele; two 3 alleles; maximum score = 9) were separated from those with at least one 4 allele conveying maximum risk for AD (one 3 allele and one 4 allele; two 4 alleles; minimum score = 12). The scale was designed based on findings that 2-carriers develop AD later and have less risk than 4-carriers, and that there is a dose-dependent effect, so that carrying two APOE 4 alleles conveys the maximum AD risk [14].

$$b_{APOE_i} = \begin{cases} 1 & if\,(count_{allele1} * count_{allele2} \succ 10) \\ 0 & otherwise \end{cases}$$ **Eq. (15)**

2.6 Information Feature, Similarity Measure, and Classification Technique

When adding all of the aforementioned bit units, we formed a binary string as follows:

b_{MMSE_i}	b_{ADAS_i}	b_{LMT_i}	b_{LHC_i}	b_{RHC_i}	b_{AGE_i}	b_{GENDER_i}	b_{SCOL_i}	b_{APOE_i}

that became our information feature. As an example here is the real data for two subjects A and B and their corresponding binary strings a and b:

Subject A

$ADAS_i$	$MMSE_i$	LMT_i	LHC_i	RHC_i	AGE_i	$GENDER_i$	$SCOL_i$	$APOE_i$
17.7	4	7	1676	1215	69	F	12	4;4
1	0	1	1	1	1	1	1	0

Subject B

$ADAS_i$	$MMSE_i$	LMT_i	LHC_i	RHC_i	AGE_i	$GENDER_i$	$SCOL_i$	$APOE_i$
2.33	3	35	2265	2474	70	M	20	4;4
0	0	0	0	0	0	0	0	0

Given those binary strings, we can compute a distance between any two subjects using an information similarity measure such as the **Hamming distance** [8]. The Hamming distance between two strings of equal length is the number of positions for which the corresponding symbols are different. It measures the minimum number of *substitutions* required to change one into the other. For binary strings a and b the Hamming distance is equal to the number of ones in a XOR b. If two words have the same length, we can count the number of digits in positions where they have different digits. The Hamming distance between the two example subjects A and B is 6. Given Hamming distances between all subjects, we used the k-nearest neighbors algorithm to classify subjects as belonging either to the CTRL or probable AD groups.

2.7 Statistical Independence Considerations

We ran all experiments in a K-fold validation setting, with K = 10 subsamples. Of the K subsamples, a single subsample (N/10 = 25 subjects on average) was retained as the validation data for testing the model, and the remaining K†−†1 subsamples were used as training data. The cross-validation process was then repeated K times, with each of the K subsamples used exactly once as the validation data. We report results average over all K folds.

Statistics and measurements were computed using the MATLAB Statistics Tool-box (The MathWorks, Natick, MA).

3 Results

3.1 Statistics and Experimental Results

Descriptive statistics and results of group tests are shown in Table 1. While group hippocampal volume differences were expected between CTRL and probable AD, the statistically significant difference in years of scolarity should be pointed out.

Three series of experiments were conducted using the methodology, primarily aimed at (a) determining the relative contribution of each bit string subgroup (core; supportive; risk) at increasing the power of the methodology for the discrimination of probable AD from CTRL; (b) combining two groups (core + supportive; core + risk; supportive + risk); and (c) computing the discriminative accuracy of the system using the complete bit string (core + supportive + risk). Accuracy, specificity, and sensitivity results are reported in Table 2. Salient results include 99.8% accuracy at the discrimination of probable AD from CTRL when using the core+supportive bit strings.

Table 1. Data statistics

	N	Age	Gender	Scol.	LHC	RHC
units		*years*		*years*	mm^3	mm^3
CTRL	138	76.3	67 F / 71 M	15.7	2123	2172
		(5.0)		(2.9)	(297)	(316)
AD	111	75.4	55 F / 56 M	14.6	1580	1638
		(7.0)		(3.1)	(325)	(374)
p-values		*N/S*	*N/S*	*0.0065*	*< 0.0001*	*< 0.0001*

Table 2. Experimental results

	Core	Supp.	Risk	Core + Supp.	Core + Risk	Supp. + Risk	Core + Supp. + Risk
Accuracy	0.983	0.752	0.580	**0.998**	0.982	0.631	0.969
Specificity	0.961	0.792	1.00	**1.00**	0.994	0.968	1.00
Sensitivity	1.00	0.708	0.04	**0.995**	0.966	0.222	0.930

4 Discussion

To our knowledge this is one of the first work that proposes a knowledge-based and explicit decision model integrating disparate data from multiple sources into a single, machine-learning framework for CTLR vs. AD discrimination. The approach is versatile and can be applied to a wide variety of problems.

Even though the choice of tests to feed the core, supportive, and risk factors has not been optimized, experimental results suggest that the current method achieves near-perfect generalization accuracy (99.8%) when following the revised criteria. It is a limitation of the current study that this argument exhibits some circularity, in that the operational definition of the diagnostic classes in the ADNI study was clinical: we are therefore confirming the original groupings. However, the accuracy of the system is remarkable given that (a) it is obtained in a leave-many-out configuration, via k-fold validation; (b) the ADNI study was not designed to test the revised criteria, but rather based its clinical diagnostic on the much wider scoped ensemble of clinical and cognitive tests (following for the most part the NINCDS-ARDA criteria); and (c) the ADNI study is widely multi-centric (>60 sites), and hence the data provided includes a significant amount of acquisition noise, relevant in clinical, cognitive, and neuroimaging results.

The accuracy reached by the system compares very favorably to the performance of other studies dealing with neuropsychological tests and hippocampal volumetry. The fact that our inclusion of so-called risk factors diminishes the performance of the core+supportive criteria is evidence to the foresight and analysis borne by the experts of the working group that authored the report [3]. However, we chose to keep these results in the current manuscript to show how such factors could be incorporated in a knowledge-based system that could be used in a different pathology.

We aim to further our work in this field by concentrating on the longitudinal validation of the approach, especially regarding the long-term confirmation of diagnostic, up to and including histopathological confirmation. Our goal is also to use this approach for the early identification of disease in prodromal AD, i.e. individuals with mild cognitive impairment.

Acknowledgments

Data collection and sharing for this project was funded by ADNI (Principal Investigator: Michael Weiner; NIH grant U01 AG024904).

References

1. Weiner, M., et al.: The Use of MRI and PET for Clinical Diagnosis of Dementia and Investigation of Cognitive Impairment: A Consensus Report. 2005. Alzheimer's Association (2005)
2. Lim, A., et al.: Clinico-neuropathological correlation of Alzheimer's disease in a community-based case series. J. Am. Geriatr. Soc. 47(5), 564–569 (1999)
3. Dubois, B., et al.: Research criteria for the diagnosis of Alzheimer's disease: revising the NINCDS-ADRDA criteria. Lancet Neurol. 6(8), 734–746 (2007)

4. von Borczyskowski, D., et al.: Evaluation of a new expert system for fully automated detection of the Alzheimer's dementia pattern in FDG PET. Nucl. Med. Commun. 27(9), 739–743 (2006)
5. Blennow, K., de Leon, M.J., Zetterberg, H.: Alzheimer's disease. Lancet 368(9533), 387–403 (2006)
6. Beinhoff, U., et al.: Gender-specificities in Alzheimer's disease and mildcognitive impairment. J. Neurol. 255(1), 117–122 (2008)
7. Koepsell, T.D., et al.: Education, cognitive function, and severity of neuropathology in Alzheimer disease. Neurology 70(19 Pt 2), 1732–1739 (2008)
8. Hamming, R.W.: Error Detecting and Error Correcting Codes. Bell System Technical Journal 26(2), 147–160 (1950)
9. Folstein, M.F., Folstein, S.E., McHugh, P.R.: Mini-mental state. A practical method for grading the cognitive state of patients for the clinician. J. Psychiatr. Res. 12(3), 189–198 (1975)
10. Rosen, W.G., Mohs, R.C., Davis, K.L.: A new rating scale for Alzheimer' s disease. Am. J. Psychiatry 141(11), 1356–1364 (1984)
11. Wechsler, D.: WMS-R Wechsler Memory Scale - Revised Manual. The Psychological Corporation, Harcourt Brace Jovanovich, Inc., New York (1987)
12. Jack, C.R.: The Alzheimer's Disease Neuroimaging Initiative (ADNI): MRI methods. J. Magn. Reson. Imaging 27(4), 685–691 (2008)
13. Hsu, Y.Y., et al.: Comparison of automated and manual MRI volumetry of hippocampus in normal aging and dementia. J. Magn. Reson. Imaging 16(3), 305–310 (2002)
14. Ohm, T.G., et al.: Apolipoprotein E isoforms and the development of low and high Braak stages of Alzheimer's disease-related lesions. Acta Neuropathol. 98(3), 273–280 (1999)

Automatic Annotation of X-Ray Images: A Study on Attribute Selection

Devrim Unay[1], Octavian Soldea[1], Ahmet Ekin[2], Mujdat Cetin[1],
and Aytul Ercil[1]

[1] Computer Vision and Pattern Analysis Laboratory, Faculty of Engineering and
Natural Sciences, Sabanci University, Turkey
unay@sabanciuniv.edu
[2] Video Processing and Analysis Group, Philips Research, The Netherlands

Abstract. Advances in the medical imaging technology has lead to an
exponential growth in the number of digital images that need to be
acquired, analyzed, classified, stored and retrieved in medical centers.
As a result, medical image classification and retrieval has recently gained
high interest in the scientific community. Despite several attempts, the
proposed solutions are still far from being sufficiently accurate for real-
life implementations.

In a previous work, performance of different feature types were investigated in a SVM-based learning framework for classification of X-Ray
images into classes corresponding to body parts and local binary patterns were observed to outperform others. In this paper, we extend that
work by exploring the effect of attribute selection on the classification
performance. Our experiments show that principal component analysis
based attribute selection manifests prediction values that are comparable to the baseline (all-features case) with considerably smaller subsets
of original features, inducing lower processing times and reduced storage
space.

1 Introduction

Storing, archiving and sharing patient information among medical centers has
become a crucial task for the medical field. Companies as well as governments are
now in anticipation of building Patient Centric IT systems such as that targeted
by the Ratu e-health project of Northern Finland[1] that focus on building a large
national electronic patient records archive.

Digital medical images, such as standard radiographs (X-Ray) and computed
tomography (CT) images, represent a huge part of the data that need to be stored
in medical centers. Manual labeling of this data is not only time consuming, but
also error-prone due to inter/intra-observer variations. In order to realize an
accurate classification one needs to develop tools that allow high performance
automatic image annotation, i.e. labeling of a given image with some text or
code without any user interaction.

---

[1] http://pre20090115.stm.fi/pr1105954774022/passthru.pdf

B. Caputo et al. (Eds.): MCBR-CDS 2009, LNCS 5853, pp. 97–109, 2010.
© Springer-Verlag Berlin Heidelberg 2010

Several attempts in the field of medical images have been performed in the past. For example, the WebMRIS system [1] aims at retrieving cervical spinal X-Ray images, whereas the ASSERT system [2] focuses on retrieving CT images of the lungs. While these efforts consider retrieving a specific body part only, other initiatives have been taken in order to retrieve multiple body parts.

The ImageCLEF Medical Image Annotation task, run as part of the Cross-Language Evaluation Forum (CLEF) campaign, is a yearly held medical image annotation challenge for automatic classification of an X-Ray image archive containing more than 10,000 images randomly taken from the medical routine. The ImageCLEF Medical Annotation dataset contains images of different body parts of people from different ages, of different genders, under varying viewing angles and with or without pathologies. Depending on the year of the challenge, participants are asked to automatically annotate these images according to classification labels that vary from 57 to 193.

A potent classification system requires the image data to be translated into a more compact and more manageable representation containing only the relevant features. Several feature representations have been investigated in the past for such a classification task. Among others, image features, such as average value over the complete image or its sub-regions [3] and color histograms [4], have been investigated. Recently in [5], texture features such as local binary patterns (LBP) [6] have been shown to outperform other types of low-level image features in classification of X-Ray images. One drawback of the mentioned work is the large number of features extracted, which may be problematic for the classification step. Retaining only the relevant features by applying attribute selection on local binary patterns, may lead to comparable classification accuracies with smaller feature sets.

Motivated by the considerations above, in this paper we explore the effect of principal component analysis based feature selection on the performance of local binary patterns applied to the ImageCLEF-2009 Medical Annotation dataset.

The paper is organized as follows. Section 2 presents our feature extraction, feature selection and classification steps in detail. Section 3, introduces the image database and the experimental evaluation process performed. Next, in Section 4, corresponding results are presented. Finally, Section 5 outlines our conclusion.

2   Method

In this work we utilize the image data from the ImageCLEF-2009 Medical Annotation task for training and testing. The architecture of the processing followed is shown in Figure 1. 12677 fully classified and unbalanced X-Ray images are available to train a classification system, which needs to be evaluated according to four different label sets including 57 to 193 distinct classes. Please note that, the data is unbalanced meaning some classes have significantly larger share among data than others.

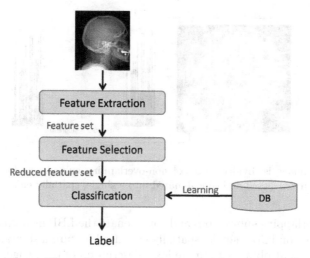

Fig. 1. Processing architecture followed

2.1   Feature Extraction

We extract *spatially enhanced local binary patterns* as features from each image
in the database. LBP [6] is a gray-scale invariant local texture descriptor with low
computational complexity. The LBP operator labels image pixels by thresholding
a neighborhood of each pixel with the center value and considering the results
as a binary number. The neighborhood is formed by a symmetric neighbor set of
P pixels on a circle of radius R. Formally, given a pixel at (x_c, y_c), the resulting
LBP code can be expressed in the decimal form as follows :

$$LBP_{P,R}(x_c, y_c) = \sum_{n=0}^{P-1} s(i_n - i_c)2^n \qquad (1)$$

where n runs over the P neighbors of the central pixel, i_c and i_n are the gray-
level values of the central pixel and the neighbor pixel, and $s(x)$ is 1 if x $\geq$ 0 and
0 otherwise.

After labeling an image with a LBP operator, a histogram of the labeled image
$f_l(x, y)$ can be defined as

$$H_i = \sum_{x,y} I(f_l(x, y) = i), \qquad i = 0, \dots, L - 1 \qquad (2)$$

where L is the number of different labels produced by the LBP operator, and
$I(A)$ is 1 if A is true and 0 otherwise.

The derived LBP histogram contains information about the distribution of
local micro-patterns, such as edges, spots and flat areas, over the image. Following [6], not all LBP codes are informative, therefore we use the uniform version
of LBP and reduce the number of informative codes from 256 to 59 (58 informative bins + one bin for noisy patterns). Following [7], we divide the images into

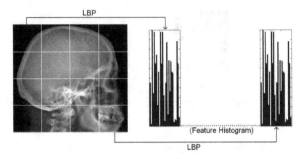

Fig. 2. The image is divided into 4x4 non-overlapping sub-regions from which LBP histograms are extracted and concatenated into a single, spatially enhanced histogram

4x4 non-overlapping sub-regions and concatenate the LBP histograms extracted from each region into a single, spatially enhanced feature histogram (Figure 2). This step aims at obtaining a more local description of the image.

Finally, we obtain a total of 944 features per image, which is a large number for the classification step. Therefore, we apply principal component analysis based feature selection.

2.2   Feature Selection: Principal Component Analysis

Principal component analysis (PCA, or Karhunen-Loéve transform) [8] is a vector space transformation often used to reduce multidimensional datasets to lower dimensions for analysis.

Given data X consisting of N samples, in PCA we first perform data normalization by subtracting the mean vector m from the data. Then the covariance matrix Σ of the normalized data $(X - m)$ is computed.

$$m = \frac{1}{N} \sum_{i=1}^{N} X_i \tag{3}$$

$$\Sigma = (X - m)(X - m)^T \tag{4}$$

Afterwards, the basis functions are obtained by solving the algebraic eigenvalue problem

$$\Lambda = \Phi^T \Sigma \Phi \tag{5}$$

where Φ is the eigenvector matrix of Σ, and Λ is the corresponding diagonal matrix of eigenvalues. Feature selection is then performed by keeping q ($q < N$) orthonormal eigenvectors corresponding to the first q largest eigenvalues of the covariance matrix. Here, q is empirically set such that total variance measured from these eigenvalues correspond to a user-defined percentage.

2.3   Classification: Support Vector Machines

SVM [9] is a popular machine learning algorithm that provide good results for general classification tasks in the computer vision and medical domains: e.g.

nine of the ten best models in ImageCLEFmed 2006 competition were based on SVM [10]. In a nutshell, SVM maps data to a higher-dimensional space using kernel functions and performs linear discrimination in that space by simultaneously minimizing the classification error and maximizing the geometric margin between the classes.

Among all available kernel functions for data mapping in SVM, Gaussian radial basis function (RBF) is the most popular choice, and therefore it is used here.

$$RBF : K(\mathbf{x}_i, \mathbf{x}_j) = exp(-\gamma \parallel \mathbf{x}_i - \mathbf{x}_j \parallel^2), \gamma > 0 \qquad (6)$$

where γ is a parameter defined by the user. Besides γ, there exists an error cost C that controls the trade-off between allowing training errors and forcing rigid margins. An optimum C value creates a soft margin while permitting some misclassifications. In this work we used LibSVM library (version 2.89) [11] for SVM and empirically found its optimum parameters (γ and C) on the dataset.

3 Experimental Setup

3.1 Image Data

The training database released for the ImageCLEF-2009 Medical Annotation task includes 12677 fully classified (2D) radiographs that are categorized into 57 classes in 2005, 119 classes in 2006 and 2007, and 193 classes in 2008. Figure 3 displays exemplary images from the database, while the distribution of the data with respect to these classes is displayed in Figure 4.

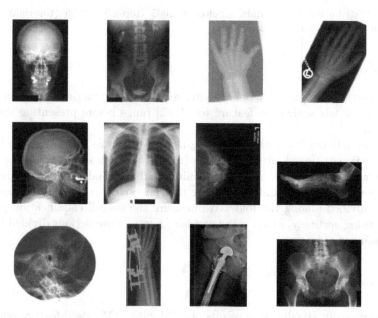

Fig. 3. Example images from the ImageCLEF-2009 Medical Annotation database

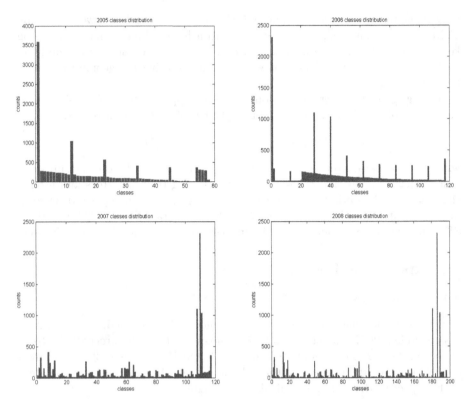

Fig. 4. Distribution of the data labeled in 2005 (top-left), 2006 (top-right), 2007 (bottom-left), and 2008 (bottom-right)

3.2 Evaluation

In order to avoid domination of attributes with greater numeric ranges over small ones, we linearly scale each feature to [-1,+1] range before presenting them to the SVM.

We evaluate our SVM-based learning using 5-fold cross validation, where the database is partitioned into five subsets (Figure 5). Each subset is used once for testing while the rest are used for training, and the final result is assigned as the average of the five validations. Note that for each validation all classes were equally divided among the folds. We measure the overall classification performance using accuracy, which is the number of correct predictions divided by the total number of images.

4   Results

We compare our classification results of SVM with PCA-based feature selection (referred to as $SVMrbf+PCA$ in the Figures 6-9) with two reference approaches: 1)

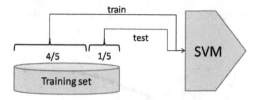

Fig. 5. Illustration of 5-fold cross validation

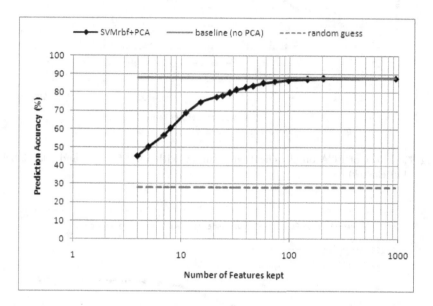

Fig. 6. Effect of PCA on classification accuracy with LBP features and 2005 labels used (57 classes)

baseline (No PCA) that refers to the SVM classification with all available features (PCA is not applied), and 2)*random guess* meaning the classifier puts all the data to the class with the highest frequency.

Figure 6 shows the effect of PCA-based feature selection on the classification accuracy of SVM for the data with 2005 labels (57 classes). Notice that when all the LBP feature set is input to the SVM (baseline case), the overall accuracy is measured as 88%, while random guess is at the level of 28%. On the other hand, with attribute selection we reach accuracy levels (87,5%) comparable to the baseline case with only about 150-200 features out of possible 944. This leads to considerable reduction in prediction time as well as storage space. This observation shows that although the used LBP^{u2} operator inherently discards non-informative patterns from the feature set, we can further refine these attributes using PCA without degrading classification accuracy.

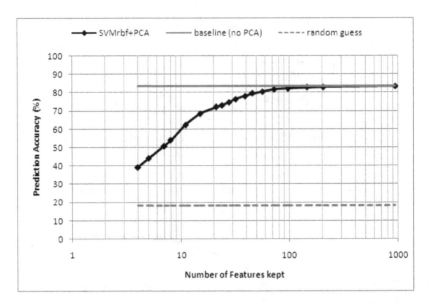

Fig. 7. Effect of PCA on classification accuracy with LBP features and 2006 labels used (119 classes)

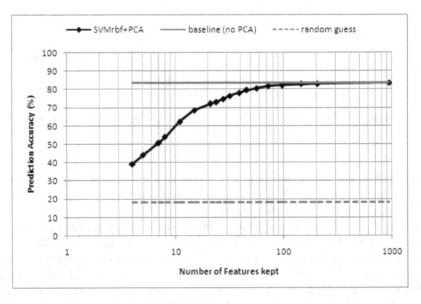

Fig. 8. Effect of PCA on classification accuracy with LBP features and 2007 labels used (119 classes)

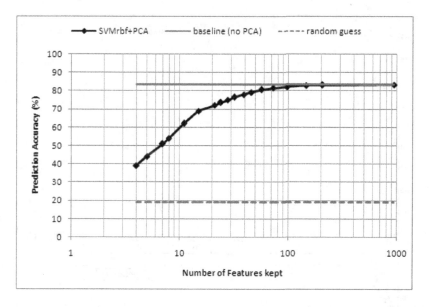

Fig. 9. Effect of PCA on classification accuracy with LBP features and 2008 labels used (196 classes)

Table 1. Computational expense of the methods on a PC with 2.13GHz processor and 2GB RAM

	Process Time (min)	Storage Space (MB)
baseline	25.5	151
with PCA	4.4	23
improvement	×5.8	×6.6

Figures 7 and 8 display the effect of PCA-based feature selection on the classification accuracy of SVM for the data with 2006 and 2007 labels (119 classes). Notice that both responses are very similar, because the mere difference between these two tasks is the replacement of 2006 labels by IRMA codes in 2007 (i.e. their number of classes are the same). In both cases we reach to observations that are consistent with the previous 2005 case (Figure 6).

Figure 9 shows the effect of PCA-based feature selection on the classification accuracy of SVM for the data with 2008 labels (193 classes). For this case, baseline accuracy is measured as 83,4%, while random guess is at the level of 18%. Similar to the previous observations, performing feature selection with PCA results in accuracy values (83%) comparable to the baseline with approximately 150-200 features. Notice that the performance levels we achieve with feature selection using 2008 labels (193 classes) are similar to those achieved by 2006 or 2007 labels (119 classes) despite the large difference in the number of classes. We believe this observation is due to the fact that manual categorizations performed

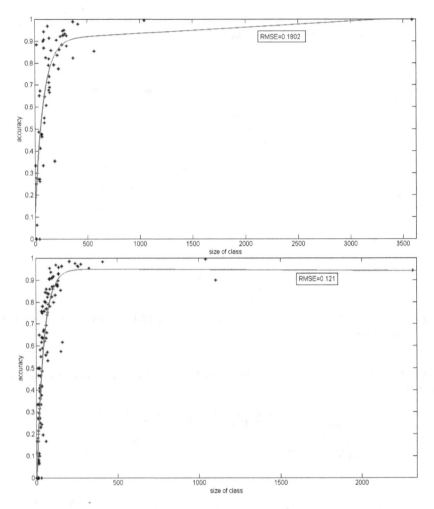

Fig. 10. Effect of class size on the classification performance for data with 2005 labels (above), and 2008 labels (below). In both plots, best fitting exponential curve to the data is displayed as a solid line and the corresponding root-mean-square-error (RMSE) is presented.

by experts for 2008 labels are more accurate and complete [12], hence more distinctive, than those carried out before.

In terms of computational expense (Table 1), the baseline approach exhibits 25.5min processing time with 151MB storage space required for the cross-validation task. On the other hand, for the proposed PCA-based approach these values are measured as 4.4min and 23MB, respectively. In consequence, the proposed PCA-based approach provides an over 5-fold improvement in processing time and storage space requirements.

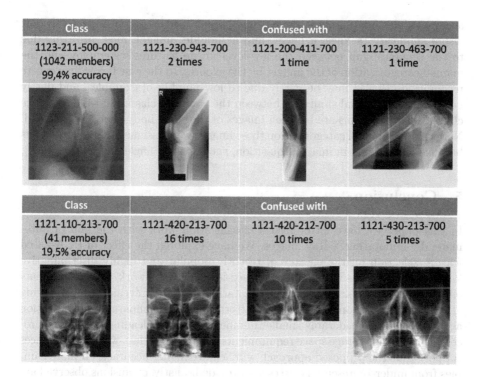

Fig. 11. Exemplary confusions realized by the proposed approach for a class with high accuracy (above), and another with relatively low accuracy (below). In each illustration, the reference class with the corresponding label, number-of-examples, accuracy, and a representative X-ray image are shown on the left, while three most-observed confusions in descending order are displayed to the right.

Detailed analysis of the class-wise performances of the proposed approach revealed that due to the imbalanced nature of the dataset classes with few samples are under-represented in the training step, and therefore their recognition is largely inaccurate. Similarly, dense classes are well represented in training, and thus our method achieves high recognition accuracies for these. Figure 10 displays the recognition performances of the proposed approach for each class with respect to the class size in the cases of data with 2005 and 2008 labels. As observed, in both cases the data is nicely represented by an exponential relationship. This observation supports the idea that the proposed approach will achieve higher accuracies if more training examples from classes with few samples are provided.

Figure 11 presents exemplary confusions realized by the proposed approach for two cases: 1)A class with high number of examples that is accurately recognized (above), and 2)A class with few samples that lead to low recognition performance (below). In the case above, class examples belong to x-ray images of abdominal region acquired from lateral view, and our approach reaches to a striking 99,4%

accuracy. The few number of confusions are from classes with relatively low visual similarity (a knee, a finger, and a shoulder image), most probably due to the semantic gap between the low-level image features we employ and the manual categorization of the images by the experts. In the case below, inaccurate recognition (19,5%) may be partly due to low number of examples, and partly because of high visual similarity between the confused classes and the reference class (Most confusions are between images of the same body part, i.e. the head. Note that, at manual categorization these images were assigned to different labels because of variation in image acquisition, such as view angle and field-of-view).

5   Conclusion

In this paper we have introduced a classification work with the aim of automatically annotating X-Ray images. We have explored the effect of PCA-based feature selection on the efficacy of recently popular and highly discriminative local binary patterns within a SVM-based learning framework. Our experiments on the ImageCLEF-2009 Medical Annotation database revealed that applying attribute selection on local binary patterns provide comparable classification accuracies with considerably smaller number of features, leading to reduced processing time and storage space requirements. Analysis on class-wise performances support that the proposed approach will achieve higher accuracies if more samples from under-represented classes are provided. Lastly, confusions observed are mainly due to three reasons: 1)semantic gap between the low-level features we use and the manual categorizations of the database, 2)presence of high visual similarity between some classes, and 3)insufficient training examples from some classes.

As future work, we plan to explore the effect of 1)using segmentation to remove background from body part regions, 2) combining different features types (e.g. shape of body part and texture), and 3)benefitting from hierarchical classification based on IRMA sub-codes.

References

1. Long, L.R., Pillemer, S.R., Lawrence, R.C., Goh, G.H., Neve, L., Thoma, G.R.: WebMIRS: web-based medical information retrieval system. In: Sethi, I.K., Jain, R.C. (eds.) Society of Photo-Optical Instrumentation Engineers (SPIE) Conference Series, December 1997, vol. 3312, pp. 392–403 (1997)
2. Shyu, C.R., Brodley, C.E., Kak, A.C., Kosaka, A., Aisen, A.M., Broderick, L.S.: Assert: a physician-in-the-loop content-based retrieval system for hrct image databases. Comput. Vis. Image Underst. 75(1-2), 111–132 (1999)
3. Rahman, M.M., Desai, B.C., Bhattacharya, P.: Medical image retrieval with probabilistic multi-class support vector machine classifiers and adaptive similarity fusion. Computerized Medical Imaging and Graphics 32(2), 95–108 (2008)
4. Mueen, A., Sapian Baba, M., Zainuddin, R.: Multilevel feature extraction and x-ray image classification. J. Applied Sciences 7(8), 1224–1229 (2007)

5. Jacquet, V., Jeanne, V., Unay, D.: Automatic detection of body parts in x-ray images. In: IEEE Computer Society Workshop on Mathematical Methods in Biomedical Image Analysis, MMBIA 2009 (2009)
6. Ojala, T., Pietikainen, M., Maenpaa, T.: Multiresolution gray-scale and rotation invariant texture classification with local binary patterns. IEEE Transactions on Pattern Analysis and Machine Intelligence 24(7), 971–987 (2002)
7. Ahonen, T., Hadid, A., Pietikäinen, M.: Face recognition with local binary patterns. In: Pajdla, T., Matas, J(G.) (eds.) ECCV 2004. LNCS, vol. 3021, pp. 469–481. Springer, Heidelberg (2004)
8. Jolliffe, I.T.: Principal Component Analysis, 2nd edn. Springer, Heidelberg (2002)
9. Burges, C.J.C.: A tutorial on support vector machines for pattern recognition. Data Mining and Knowledge Discovery 2(2), 121–167 (1998)
10. Müller, H., Deselaers, T., Deserno, T., Clough, P., Kim, E., Hersh, W.: Overview of the imageclefmed 2006 medical retrieval and medical annotation tasks. In: Peters, C., Clough, P., Gey, F.C., Karlgren, J., Magnini, B., Oard, D.W., de Rijke, M., Stempfhuber, M. (eds.) CLEF 2006. LNCS, vol. 4730, pp. 595–608. Springer, Heidelberg (2007)
11. Chang, C.C., Lin, C.J.: LIBSVM: a library for support vector machines (2001), http://www.csie.ntu.edu.tw/~cjlin/libsvm
12. Tomasi, T., Caputo, B., Welter, P., G/"ult, M.O., Deserno, T.M.: Overview of the clef 2009 medical image annotation track. In: CLEF Working Notes 2009 (2009)

Multi-modal Query Expansion Based on Local Analysis for Medical Image Retrieval

Md. Mahmudur Rahman, Sameer K. Antani, Rodney L. Long,
Dina Demner-Fushman, and George R. Thoma

U.S. National Library of Medicine,
National Institutes of Health, Bethesda, MD, USA
{rahmanmm,santani,rlong,ddemner,gthoma}@mail.nih.gov

Abstract. A unified medical image retrieval framework integrating visual and text keywords using a novel multi-modal query expansion (QE) is presented. For the content-based image search, visual keywords are modeled using support vector machine (SVM)-based classification of local color and texture patches from image regions. For the text-based search, keywords from the associated annotations are extracted and indexed. The correlations between the keywords in both the visual and text feature spaces are analyzed for QE by considering local feedback information. The QE approach can propagate user perceived semantics from one modality to another and improve retrieval effectiveness when combined in multi-modal search. An evaluation of the method on imageCLEFmed'08 dataset and topics results in a mean average precision (MAP) score of 0.15 over comparable searches without QE or using only single modality.

1 Introduction

Medical image retrieval based on multi-modal sources has been recently gaining popularity due the large amount of text-based clinical data available in the form of case and lab reports. Improvement in retrieval performances has been noted by fusing evidence from the textual information and the visual image content in a single framework. The results of the past ImageCLEFmed [1] tracks suggest that the combination of visual and text based image searches provides better results than using the two different approaches individually [1]. While there is a substantial amount of completed and ongoing research in both the text and content based image retrieval (CBIR) in medical domain [2,3], much remains to be done to see how effectively these two approaches can complement each other in an integrated interactive framework based on query reformulation.

To increase the effectiveness and reduce the ambiguity due to the word mismatch problem in text information retrieval, a variety of query reformulation strategies based on term co-occurrence or term similarity have been investigated [4,5,6,7]. These techniques exploit term (keyword) dependency as term clustering

---

[1] http://imageclef.org/2008/medical

B. Caputo et al. (Eds.): MCBR-CDS 2009, LNCS 5853, pp. 110–119, 2010.
© Springer-Verlag Berlin Heidelberg 2010

in document collection based on either global or local analysis [5]. In a global analysis, all documents in the collection are analyzed to determine a global thesaurus-like structure that defines term relationships. This structure is then utilized to select additional terms for QE. In local analysis, the top retrieved documents for a query are examined at query time without any assistance from the user, in general to determine the terms for QE.

On the other hand, due to the nature of the low-level continuous feature representation in majority of the CBIR systems [8,9], the idea of QE cannot be directly applied and is relatively new in this domain [10,11]. For example, a context expansion approach has been recently explored in [10] by expanding the key regions of the (image) queries using highly correlated environmental regions according to an image thesaurus. In [11], the authors attempt to automatically annotate and retrieve images by applying QE in its relevance model based on a set of training images. Here, images are modeled with a *bag-of-concepts* (e.g., *bag-of-words* in text) approach of vector space model (VSM) in information retrieval [12]. These approaches are either data dependent over the entire collection or dependent on the associated keywords. In this paper, we explore a fundamentally different QE technique in a unified multi-modal framework, which is based on the correlation analysis of both visual and text keywords and relies only on the local feedback information. The aim of this framework is to develop methods that can combine and take advantage of complementary information from both modalities through application of cross-modal QE mechanism.

The proposed approach and an evaluation of its efficacy are presented as follows: in Section 3, we briefly describe the image representation approach in visual and text keyword spaces. In Section 4, we describe the proposed query expansion strategy based on local analysis. The experiments and analysis of the results are presented in Section 5 and finally Section 6 provides the conclusions.

2   Image Representation in Visual Keyword Space

In a heterogeneous collection of medical images, it is possible to identify specific local patches that are perceptually and/or semantically distinguishable, such as homogeneous texture patterns in grey level radiological images, differential color and texture structures in microscopic pathology and dermoscopic images, etc. The variation in these local patches can be effectively modeled as visual keywords by using supervised learning based classification techniques, such as the support vector machine (SVM) [13]. In its basic formulation, the SVM is a binary classification method that constructs a decision surface and maximizing the inter-class boundary between the samples. A number of methods have been proposed for multi-class classification by solving many two-class problems and combining their predictions. For visual keyword generation, we utilize one such voting-based multi-class SVM known as *one-against-one* or pairwise coupling (PWC) [14].

In order to perform the learning, a set of L labels are assigned as $C = \{c_1, \cdots, c_i, \cdots, c_L\}$, where each $c_i \in C$ characterizes a visual keyword. The training set of the local patches that are generated by a fixed-partition based approach and represented by a combination of color and texture moment-based features. For SVM training, the initial input to the system is the feature vector set of the patches along with their manually assigned corresponding concept labels. Images in the data set are annotated with visual keyword labels by fixed partitioning each image I_j into l regions as $\{\mathbf{x}_{1_j}, \cdots, \mathbf{x}_{k_j}, \cdots, \mathbf{x}_{l_j}\}$, where each $\mathbf{x}_{k_j} \in \Re^d$ is a combined color and texture feature vector. For each $\mathbf{x}_{k_j}$, the visual keyword probabilities are determined by the prediction of the multi-class SVMs as [14]

$$p_{ik_j} = P(y = i \mid \mathbf{x}_{k_j}), \ 1 \le i \le L. \tag{1}$$

Finally, the category label of x_{k_j} is determined as c_m, which is the label of the category with the maximum probability score. Hence, the entire image is thus represented as a two-dimensional index linked to the visual keyword labels. Based on this encoding scheme, an image I_j is represented as a vector of weighted visual keywords as

$$\mathbf{f}_j^I = [w_{1_j}, \cdots, w_{i_j}, \cdots w_{L_j}]^{\mathrm{T}} \tag{2}$$

where each w_{i_j} denotes the weight a visual keyword $c_i, 1 \le i \le L$ in image I_j, depending on its information content. The popular "tf-idf" term-weighting scheme [12] is used in this work, where the element w_{i_j} is expressed as the product of local and global weights.

3   Image Representation in Text Keyword Space

For the text-based image search, it is necessary to transform the annotation files in XML formats into an easily accessible representation known as the *index*. In this case, information from only relevant tags are extracted and preprocessed by removing stop words that are considered to be of no importance for the actual retrieval process. Subsequently, the remaining words are reduced to their stems, which finally form the set $T = \{t_1, t_2, \cdots, t_N\}$ of index terms or keywords of the annotation files. Next, the annotation files (document) are modeled as a vector of keywords as

$$\mathbf{f}_j^D = [\hat{w}_{1_j}, \cdots, \hat{w}_{i_j}, \cdots \hat{w}_{N_j}]^{\mathrm{T}} \tag{3}$$

where each $\hat{w}_{i_j}$ denotes the weight of a keyword $t_i, 1 \le i \le N$ in the annotation of image I_j. A weighting scheme has two components: a global weight and a local weight. The global importance of a term is indicating its overall importance in the entire collection, weighting all occurrences of the term with the same value. The popular *tf-idf* term-weighting scheme [12] is used in this work, where the local weight is denoted as $L_{ji} = log(f_{ji}) + 1$, f_{ji} is the frequency of occurrence of keyword t_i in document D_j. The global weight G_i is denoted as inverse document frequency as $G_i = log(M/M_i)$, for $i = (1, \cdots,, N)$, where M_i be the number of documents in which t_i is found and M is the total number of documents in the collection. Finally, the element $\hat{w}_{i_j}$ is expressed as the product of local and global weight as $\hat{w}_{i_j} = L_{ji} * G_i$. This weighting scheme amplifies the influence of terms, which occur often in a document (e.g., *tf* factor), but relative rarely in the whole collection of documents (e.g., *idf* factor[12]. A query D_q is also represented as a vector of length N as $\mathbf{f}_q^D = [\hat{w}_{1_q}, \cdots, \hat{w}_{i_q}, \cdots, \hat{w}_{N_q}]^T]$, and the similarity (cosine) of the document vectors to a query vector gives a retrieval score to each document, allowing comparison and ranking of documents.

4   Multi-modal QE Based on Local Analysis

Query expansion based on local feedback and cluster analysis has been one of the most effective methods for expanding queries in text retrieval domain [4,5,7]. Generally, this approach expands a query based on the information from the top retrieved documents for that query without any assistance from the user. The correlated terms are identified and ranked in order of their potential contribution to the query and are re-weighted and appended to the query [4]. Before presenting our query expansion method, some basic terminologies need to be defined as follows:

Definition 1. *Let us consider q as a multi-modal query, which has an image part as I_q and a text part as D_q. The similarity between q and a multi-modal item j, which also has also two parts (e.g., image I_j and context D_j), is defined as*

$$Sim(q, j) = \omega_I Sim_I(I_q, I_j) + \omega_D Sim_D(D_q, D_j) \tag{4}$$

Here, ω_I and ω_D are normalized inter-modality weights within the text and image feature spaces. In this framework, the individual image $Sim_I(I_q, I_j)$ and text $Sim_D(D_q, D_j)$ based similarities are computed based on the Cosine distance measure [12].

Definition 2. *For the given query q, the set S_l of retrieved images along with associated annotations is called the* local image set. *Also, the set $C_l \subseteq C$ of all distinct visual keywords $c_i \in C_l$ and the set $T_l \subseteq T$ of all distinct text keywords $t_i \in T_l$ in the local image set S_l is called the* local vocabulary of visual and text keywords *respectively.*

Since, the correlated terms for an expansion are those present in the local cluster, we first need to generate such clusters from S_l and thereafter from C_l and T_l. To generate the cluster, we rely on a local correlation matrix that is built based on the co-occurrence of keywords inside images and associated annotations. Let $\mathbf{A}_I^I = [a_{uv}]$ be a $|C_l| \times |C_l|$ matrix in which the rows and columns are associated with the visual keywords in C_l. Each entry a_{uv} expresses a normalized correlation factor between visual keywords c_u and c_v as

$$a_{uv} = n_{uv}/(n_u + n_v - n_{uv}) \tag{5}$$

where n_u be the number of images in S_l that contain the keyword c_u, n_v be the number of images that contain the keyword c_v, and n_{uv} be the number of the top retrieved images in S_l that contain both keywords. The entry a_{uv} measures the ratio between the number of images where both c_u and c_v appear and the

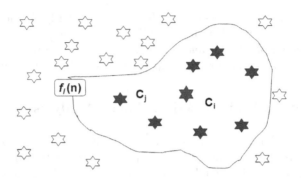

Fig. 1. Visual keyword c_j as a neighbor of the keyword c_i based on a local cluster

total number of images in S_l where either c_u or c_v appear and its value ranges to $0 \leq a_{uv} \leq 1$. If c_u and c_v have many co-occurrences in images, the value of a_{uv} increases and the images are considered to be more correlated. In a similar fashion, we can generate a $|D_l| \times |D_l|$ matrix $\mathbf{A}_l^D$ in which the rows and columns are associated with the keywords in T_l. The global version of this matrix, which is termed as the *connection matrix*, is utilized in a fuzzy information retrieval approach in [16]. By generating the above matrices, we can use them to build local correlation clusters for the multi-modal query expansion. Let $f_i(n)$ be a function that takes the i-th row and return the ordered set of n largest values

---

Algorithm 1. Query Expansion through Local Analysis

1: Initialize a temporary expanded query vector of query image I_q as $\mathbf{f}_q^e = [\hat{w}_{1_q} \ \hat{w}_{2_q} \cdots \hat{w}_{iq} \cdots \hat{w}_{L_q}]^T$ where each $\hat{w}_{i_q} = 0$.
2: For an original query vector $\mathbf{f}_q^o = [w_{1_q} \ w_{2_q} \cdots w_{i_q} \cdots w_{L_q}]^T$ of I_q, perform initial retrieval.
3: Consider, the top ranked K images as the local image set S_l and construct the local correlation matrix $\mathbf{A}_l^I$ based on equation (5)
4: **for** $i = 1$ to L **do**
5:    **if** $w_{i_q} > 0$ **then**
6:       Consider the i-th row in $\mathbf{A}_l^I$ for the visual keyword c_i.
7:       Return $f_i(n)$, the ordered set of n largest values m_{ij}, where $i \neq j$, therefore $c_j \in C_l - \{c_i\}$.
8:       **for** each c_j **do**
9:          Add and re-weight the corresponding element in query vector as $\hat{w}_{j_q} + = w_{i_q} - ((w_{i_q} - 0.1) \times k/n)$, where k is the position of c_j in the rank order.
10:       **end for**
11:    **end if**
12: **end for**
13: Obtain the re-formulated or modified query vector as $\mathbf{f}_q^m = \mathbf{f}_q^e + \mathbf{f}_q^o$.
14: Perform the image-based search with the modified query vector $\mathbf{f}_q^m$.
15: Continue the process, i.e., steps 3 to 14 until no more changes are noticed.

---

a_{ij} from $\mathbf{A}_l^I$ ($\mathbf{A}_l^D$), where j varies over the set of keywords and $i \neq j$. Then $f_i(n)$ defines a local correlation cluster around the visual keyword c_i (t_i) as shown as a blue star in Figure 1). Here, the concept c_j (t_j) is located within a neighborhood $f_i(n)$ associated with the concept c_i (t_i) as shown as the red stars in Figure 1).

Now, the keywords that belong to clusters associated with the query q can be used to expand the original query. Often these neighbor concepts are correlated by the current query context [12]. The steps of the query expansion process for visual keywords based on the correlation cluster on $\mathbf{A}_l^I$ are given in Algorithm 1. Similar steps are also applied for the textual query expansion based on the correlation cluster on $\mathbf{A}_l^D$. Based on the Step 9 of the Algorithm 1, weights are assigned in such a way that a top ranked keyword in a ordered set gets the largest weight value than the next one in the set. After this expansion process, new keywords may have been added to the original query based on the Step 9 and the weight of an original query concept may have been modified had the concept belonged to the top ranked concepts based on the Step 13 of the algorithm.

Figure 2 shows the process flow diagram of the above multi-modal query expansion-based search approach. Here, the top-middle portion shows how a search is initiated simultaneously based on both text and image parts of a

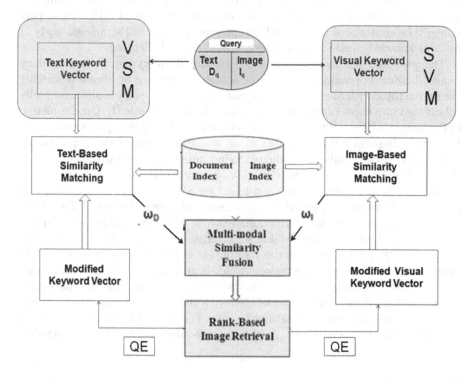

Fig. 2. Process flow diagram of the multimodal query expansion framework

multimodal query and later the individual results are combined for a final ranked result list. By observing the result list, a user can provide local feedback information for query expansion in both the text and visual feature spaces. In the next iteration, user will get a new result list based on the modified query vectors. Since, a query can be represented with both visual and text keywords, it can be initiated either by a keyword-based search or by a image-query-by-example (QBE) search. The query expansion approach can be used to automatically expand a textual query using related keywords obtained from top retrieved (based on user's feedback) annotation files of associated images. In a similar fashion, the visual QBE can be reformulated using visual keywords from top retrieved relevant images based on a CBIR or text search in the previous iteration. The flexibility in such a search process implicitly creates a semantic network that links text keywords with visual keywords and vice versa.

5   Experiments and Results

To evaluate the retrieval effectiveness, experiments are performed in a benchmark medical image collection from ImageCLEFmed'08 [1]. This collection contains more than 67,000 images of different modalities from the RSNA journals[2] Radiology and Radiographics. For each image, the text of the figure caption is supplied as free text. In some cases, however, the caption is associated with a multi-part image. The contents of this collection represent a broad and significant body of medical knowledge, which makes the retrieval more challenging.

The proposed methods are evaluated on the 30 query topics developed by imageCLEFmed organizers. Each topic is a short sentence or phrase describing the search topic with one to three "relevant" images. The query topics are equally subdivided into three categories: visual, mixed, and semantic [1]. On completion of the imageCLEFmed'08 task, a set of relevant results for all topics were created by considering top retrieval results of all submitted runs of the participating groups. Retrieval results are evaluated using uninterpolated (arithmetic) *Mean Average Precisions (MAP)* and *Precision* at rank 20 (P 20).

For the visual keyword generation based on SVM learning, 30 local concept categories are manually defined, such as tissues of lung or brain of CT or MRI, bone of chest, hand, or knee X-ray, microscopic blood or muscle cells, dark or white background, etc. The training set consists of less then 1% images of the entire collection. Each image in the training set is partitioned into an 8×8 grid generating 64 non-overlapping regions, which is proved to be effective to generate the local patches. Only the regions that conform to at least 80% of a particular concept category are selected and labeled with the corresponding category label due to the consideration of robustness to noise. For the SVM training, we utilized the radial basis function (RBF). A 10-fold cross-validation (CV) is conducted to

---

[2] http://www.rsnajnls.org

Table 1. Retrieval results for different methods

Method	Modality	QE	MAP	P20
Visual Keyword	Image	Without	0.025	0.0717
Visual Keyword-QE	Image	With	0.028	0.0767
Text Keyword	Text	Without	0.1253	0.1469
Text Keyword-QE	Text	With	0.1311	0.1491
Visual-Text Keyword	Image+Text	Without	0.1426	0.1522
Visual-Text Keyword-QE	Image+Text	With	0.1501	0.1564

find the best tunable parameters C and γ of the RBF kernel. After finding the best values of the parameters $C = 200$ and $\gamma = 0.02$ of the RBF kernel with a CV accuracy of 81.01%, they are utilized for the final training to generate the SVM model file. We utilized the *LIBSVM* software package [15] for implementing the multi-class SVM classifier. For text based indexing, we only consider the keywords (after removing stop words and stemming) from the *"ArticleTitle"* tag of the XML formats of each abstract, which are linked by *one-to-one* or *one-to-many* relationship with images in the collection.

The performances are compared with and without using any QE in different feature spaces, i.e., visual, text, and multi-modal as shown in Table 1. For the automatic simulation of QE, we considered top 20 retrieved images from the previous iteration as the local feedback for the next iteration and selected three additional visual and text keywords from the local clusters for each query keywords in both the visual and text feature spaces. For multi-modal retrieval, the search is initiated simultaneously based on both text and image parts of a query and later the individual results are linearly combined (with weight $\omega_I = 0.3$ and $\omega_D = 0.7$) for a final ranked result list.

It is clear from Table 1 that the retrieval performance was improved for the QE-based approaches (even after using only one iteration of feedback) compared to the case when no QE is utilized with image representation in visual and text keyword spaces. The proposed QE method performed well in all cases, i.e., whether it was applied to a single modality or was applied in the multi-modal search with a linear combination scheme. In general, we achieved around 4-5% increase in MAP scores for all QE based searches compared to searches without any expansion. For example, for the search with multi-modal QE (e.g., Visual-Text Keyword-QE) where visual and textual expansions are performed together, we achieved the best MAP score of 0.15. Finally, from the results, we can conjecture that there exists enough correlation between the visual-visual and text-text keywords, which can be exploited with QE or modification process. It is also evident that combining both the visual and text keyword-based features as well as using QE techniques can significantly improve retrieval performance. A comparison with results from approaches by other imageCLEFmed'08 participants is not possible due to lack of evidence on their use of any query-expansion methods or kinds of relevance feedback techniques in a multi-modal framework.

6   Conclusions

This paper investigates a novel multimodal query-expansion (QE) technique for medical image retrieval inspired by approaches in text information retrieval. The proposed technique exploits correlations between the visual and text keywords using a local analysis approach. We observe that there exists enough correlation between keywords within each modality and exploiting this property reduces the keyword mismatch problem. Furthermore, a standard image dataset has provided enough reliability for objective performance evaluation that demonstrates the efficacy of the proposed method.

Acknowledgment

This research is supported by the Intramural Research Program of the National Institutes of Health (NIH), National Library of Medicine (NLM), and Lister Hill National Center for Biomedical Communications (LHNCBC). We would like to thank the CLEF [1] organizers for making the database available for the experiments.

References

1. Müller, H., Müller, H., Cramer, J.K., Kahn, C.E., Hatt, W., Bedrick, S., Hersh, W.: Overview of the ImageCLEFmed 2008 medical image retrieval task. In: 9th Workshop of the Cross-Language Evaluation Forum, CLEF 2008, Aarhus, Denmark (2008)
2. Müller, H., Michoux, N., Bandon, D., Geissbuhler, A.: A review of content-based image retrieval applications–clinical benefits and future directions. International Journal of Medical Informatics 73, 1–23 (2004)
3. Wong, T.C.: Medical Image Databases. Springer, New York (1998)
4. Xu, J., Croft, W.B.: Improving the effectiveness of information retrieval with local context analysis. ACM Trans. on Info. Sys. 18(1), 79–112 (2000)
5. Xu, J., Croft, W.B.: Query Expansion Using Local and Global Document Analysis. In: Proc. 19th Annual Int'l ACM SIGIR Conf. on Research and Develop. in Info. Retrieval, pp. 4–11 (1996)
6. Crouch, C.J.: An approach to the automatic construction of global thesauri. Info. Process. and Management 26(5), 629–640 (1990)
7. Attar, R., Fraenkel, A.S.: Local feedback in full-text retrieval systems. Journal of the ACM (J. ACM) 24(3), 397–417 (1977)
8. Liua, Y., Zhang, D., Lu, G., Ma, W.Y.: A survey of content-based image retrieval with high-level semantics. Pattern Recognition 40(1), 262–282 (2007)
9. Datta, R., Joshi, D., Li, J., Wang, J.Z.: Image retrieval: Ideas, influences, and trends of the new age. ACM Computing Surveys 40(2), 1–60 (2008)
10. Wang, X.J., Ma, W.Y., Li, X.: Exploring Statistical Correlations for Image Retrieval. Multimedia Systems 11(4), 340–351 (2006)
11. Jeon, J., Lavrenko, V., Manmatha, R.: Automatic Image Annotation and Retrieval using Cross-Media Relevance Models. In: Proc. of the 26th annual international ACM SIGIR conference on Research and development in informaion retrieval, pp. 119–126 (2003)

12. Baeza-Yates, R., Ribiero-Neto, B.: Modern Information Retrieval. Addison Wesley, Reading (1999)
13. Vapnik, V.: Statistical Learning Theory. Wiley, New York (1998)
14. Ting-Fan, W., Chih-Jen, L., Ruby, C.W.: Probability Estimates for Multi-class Classification by Pairwise Coupling. J. of Machine Learning Research 5, 975–1005 (2004)
15. Chang, C.C., Lin, C.J.: LIBSVM: a library for support vector machines (2001), http://www.csie.ntu.edu.tw/~cjlin/libsvm
16. Yasushi, O., Tetsuya, M., Kiyohiko, K.: A fuzzy document retrieval system using the keyword connection matrix and a learning method. Fuzzy Sets and Systems 39(2), 163–179 (1991)

Author Index